NASA AERONAUTICS BO

Coming Home

Reentry and Recovery from Space

NASA

Roger D. Launius and Dennis R. Jenkins

Table of Contents

Acknowledgments

As with any published work, the authors of this book inevitably incurred debts to patrons, colleagues, and fellow travelers. In writing *Coming Home: Reentry and Recovery from Space*, we certainly relied on the help of many people. Most importantly, this project was made possible through a grant from the NASA Aeronautics Mission Directorate, where Tony Springer has been instrumental in sponsoring a large number of important historical studies. We are indebted to the many peer reviewers who offered important suggestions for improving the manuscript. We hope that they will be pleased with the final product.

The authors would like to acknowledge several individuals who aided in the preparation of this book. Only through the assistance of these key people were we able to assemble the materials for this volume in some semblance of order. For their many contributions to this project, we especially wish to thank Jane Odom and her staff of archivists at the NASA History Program Office for help-ing to track down information and correct inconsistencies. We would also like to thank Steve Dick, Bill Barry, Steve Garber, and Nadine Andreassen at NASA; the staffs of the NASA Headquarters Library and the Scientific and Technical Information Program who provided assistance in locating materials; Marilyn Graskowiak and her staff at the National Air and Space Museum (NASM) Archives; and the many archivists and scholars throughout NASA and other organizations whose work was used to complete this manuscript. Archivists at the National Archives and Records Administration were also instrumental in making this study possible.

Patricia Graboske, head of publications at NASM, provided important guidance for this project. Thanks to Tom Watters, Bruce Campbell, John Grant, and Jim Zimbelman of NASM's Center for Earth and Planetary Studies for information on science missions. We also thank Karen McNamara for information about the Genesis and Stardust missions.

In addition to these individuals, we wish to acknowledge the following people who aided in a variety of ways: Debbora Battaglia, David Brandt, William E. Burrows, Erik Conway, Bob Craddock, Tom D. Crouch, Gen. John R. Dailey, David H. DeVorkin, Robert Farquhar, Jens Feeley, James Rodger Fleming, James Garvin, Lori B. Garver, G. Michael Green, Barton C. Hacker, James R. Hansen, Wes Huntress, Peter Jakab, Violet Jones-Bruce, Sylvia K. Kraemer, John Krige, Alan M. Ladwig, W. Henry Lambright, Jennifer Levasseur, John M. Logsdon, W. Patrick McCray, Howard E. McCurdy, Jonathan C. McDowell,

Ted Maxwell, Valerie Neal, Allan A. Needell, Michael J. Neufeld, Frederick I. Ordway III, Scott Pace, Robert Poole, Alan Stern, Harley Thronson, and Margaret Weitekamp. All of these people would disagree with some of this book's observations, but such is the boon and bane of scholarly inquiry.

Introduction

Saturday morning, February 1, 2003, dawned as most other weekend days in America in the early 21st century. People went about their morning routines and perhaps turned on their radios or televisions to catch the latest news. Those that did so learned of a tragic story that captured the Nation's attention throughout that day as well as the days that followed. At 8:51:14 eastern standard time (EST), during what most people believed to be a routine landing of the Space Shuttle Columbia at the Kennedy Space Center, FL, the vehicle started breaking apart. By 8:58:19 EST, a large-scale wing deformation had taken place, and a fraction of second later Mission Control at the Johnson Space Center in Houston, TX, lost all telemetry. Thereafter, the main vehicle disintegrated at 9:00:18 EST, based on video imagery. In the process, a crew of seven astronauts lost their lives; debris from the Shuttle scattered across the American South, from the Pacific coast to the Gulf states; the National Aeronautics and Space Administration (NASA) grounded the Shuttle fleet pending an investigation; and as the Nation grieved, many in technical communities worked to determine what had taken place and, most importantly, why it had happened.

The subsequent Columbia Accident Investigation Board (CAIB), under the leadership of Retired Admiral Harold W. Gehman, Jr., identified the technical cause of the accident as taking place during launch when the External Tank's foam insulation struck the leading edge of the left wing and damaged the thermal protection system. Equally important, the CAIB explored the non-technical managerial, organizational, and cultural issues that had allowed the technical problems that led to the wing breech to go unchecked for the life of the Shuttle program. At a fundamental level, those larger issues emerged as more significant than the technical factors leading to the Columbia accident.

The tragic loss of Columbia brought to the fore one of the most difficult issues NASA has had to deal with since the beginning of space flight: how its space systems operate while transiting the atmosphere as they return to Earth. The technologies for the reentry and recovery from space might change over time, but the challenge remains one of the most important and vexing in the rigorous efforts to bring spacecraft and their crews and cargo home successfully. Returning to Earth after a flight into space is a fundamental challenge, and contributions from the NASA Aeronautics Research Mission Directorate in aerodynamics, thermal protection, guidance and control, stability, propulsion, and landing systems have proven critical to the success of the human space flight and other space programs. Without this base of fundamental and applied research, the capability to fly into space would not exist.

Accordingly, this study represents a means of highlighting the myriad of technological developments that made possible the safe reentry and return from space and the landing on Earth. This story extends back at least to the work of Walter Hohmann and Eugen Sänger in Germany in the 1920s and involved numerous aerospace engineers at the National Advisory Committee for Aeronautics (NACA)/NASA Langley and the Lewis (now the John H. Glenn Research Center at Lewis Field) and Ames Research Centers. For example, researchers such as H. Julian Allen and Alfred J. Eggers, Jr., at Ames pioneered blunt-body reentry techniques and ablative thermal protection systems in the 1950s, while Francis M. Rogallo at Langley developed creative parasail concepts that informed the development of the recovery systems of numerous reentry vehicles.

The chapters that follow relate in a chronological manner the way in which NASA has approached the challenge of reentering the atmosphere after a space mission and the technologies associated with safely dealing with the friction of this encounter and the methods used for landing safely on Earth. The first chapter explores the conceptual efforts to understand the nature of flight to and from space and the major developments in the technologies of reentry and landing that took place before the beginning of the space age in 1957.

While most proposals for satellites prior to 1957 avoided the difficult problem of reentry, the work done on solving the practical problems of space flight made it increasingly obvious that returning to Earth represented a major step in flight. Beginning with theoretical work, experimentation followed to help understand the heating of reentry during ballistic missile development in the 1950s.

Three reentry approaches dominated thinking at the time. The first was a heat-sink concept that sought to move quickly from space through the upper atmosphere. The heat sink, usually a large mass of copper or beryllium, would simply absorb the heat as the object plummeted through the atmosphere. However, superheating proved a serious problem as range and speed grew, and engineers soon realized that heat sinks were unacceptable for orbital reentry. The second approach, championed by Wernher von Braun and his rocket team in Huntsville, AL—and championed by post–World War II German officer Walter R. Dornberger at Bell Aircraft—called for circulating a fluid through the spacecraft's skin to soak up the heat of reentry. Von Braun's grandiose vision foresaw astronauts returning from wheeled space stations aboard huge spaceplanes, but when challenged to develop actual hardware, he realized that there was no way for the heat to be absorbed without killing the occupants. For orbital flight, both of these concepts gave way to Allen and Eggers's blunt-body concept, which fundamentally shaped the course of space flight research and provided the basis for all successful reentry vehicles. At the same time, John V. Becker and others at the NACA's Langley Research Center were championing the X-15, a winged research airplane intended to demonstrate the ability to fly back from near-space to normal runway landings. Ironically, the X-15 ultimately used a variation of Allen's blunt-body theory.

The X-planes of the 1950s, especially the X-15, proved critical to advancing reentry and recovery technologies for space flight and are the subject of the second chapter in this study. They advanced knowledge of the blunt-body reentry design with an ablative shield to deal with the heat generated by atmospheric friction. This approach proved significant for piloted missions, but it has also been used successfully in reconnaissance, warhead, and scientific reentry from the 1950s to the present day. Additionally, the question of what materials to use to protect the spacecraft during blunt-body reentry emerged, and research on metallic, ceramic, and ablative heat shields prompted the decision to employ ablative technology. All of these decisions required tradeoffs, and the story of the process whereby these decisions were made and implemented offers an object lesson for current engineers involved in making difficult technical choices.

Chapter 2 also investigates the methods of landing once a spacecraft reaches subsonic speeds. Once the orbital energy is converted and the heat of reentry dissipated, the spacecraft must still be landed gently in the ocean or on land. Virtually all of the early concepts for human space flight involve spaceplanes that flew on wings to a runway landing; Sänger's antipodal bomber of the 1940s did so as did von Braun's popular concepts. However, these proved impractical for launch vehicles available during the 1950s, and capsule concepts that returned to Earth via parachute proliferated largely because they represented the "art of the possible" at the time.

Chapter 3 tells the story of reentry from space and landing on Earth from the beginning of the space age through the end of the Apollo program. During that period, NASA and other agencies concerned with the subject developed capsules with blunt-body ablative heat shields and recovery systems that relied on parachutes. The Department of Defense (DOD) tested this reentry concept publicly with Project SCORE (Signal Communication by Orbiting Relay Equipment) in 1958 and employed it throughout the CORONA satellite reconnaissance program of the 1960s, snatching in midair return capsules containing unprocessed surveillance footage dangling beneath parachutes. With the Mercury program, astronauts rode a blunt-body capsule with an ablative heat shield to a water landing, where the Navy rescued them. Project Gemini eventually used a similar approach, but NASA engineers experimented with a Rogallo wing and a proposed landing at the Flight Research Center (now Dryden Flight

Research Center) on skids similar to those employed on the X-15. When the Rogallo wing failed to make the rapid progress required, NASA returned to the parachute concept used in Mercury and essentially used the same approach in Apollo, although with greatly improved ablative heat shields.

At the same time, the DOD pursued a spaceplane concept with the X-20 Dyna-Soar orbital vehicle that would have replaced the ablative heat shield with a reusable metallic heat shield and a lifting reentry that allowed the pilot to fly the vehicle to a runway landing. This is also the general approach pursued by the DOD with its Aerothermodynamic Elastic Structural Systems Environmental Tests (ASSET) and Martin X-23A Precision Reentry Including Maneuvering reEntry (PRIME) vehicles. NASA and DOD also experimented with lifting body concepts. Engineers were able to make both of those approaches to reentry and landing work, making tradeoffs on various other capabilities in the process. The eventual direction of these programs was influenced more by technological choices than by obvious decisions.

Even as Apollo was reaching fruition in the late 1960s, NASA made the decision to abandon blunt-body capsules with ablative heat shields and recovery systems that relied on parachutes for its human space flight program. Instead, as shown in chapters 4 and 5, it chose to build the Space Shuttle, a winged reusable vehicle that still had a blunt-body configuration but used a new ceramic tile and reinforced carbon-carbon for its thermal protection system. Parachutes were also jettisoned in favor of a delta-wing aerodynamic concept that allowed runway landings. Despite many challenges and the loss of one vehicle and its crew due to a failure with the thermal protection system, this approach has worked relatively effectively since first flown in 1981. Although NASA engineers debated the necessity of including jet engines on the Shuttle, it employed the unpowered landing concept demonstrated by the X-15 and lifting body programs at the Flight Research Center during the 1960s. These chapters lay out that effort and what it has meant for returning from space and landing on Earth.

The concluding chapter explores efforts to develop new reentry and landing concepts in the 1990s and beyond. During this period, a series of ideas emerged on reentry and landing concepts, including the return of a metallic heat shield for the National Aero-Space Plane and the X-33, the Roton rotary rocket, the DC-X powered landing concept, and the Crew Exploration Vehicle (CEV) of the Constellation program between 2005 and 2009. In every case, these projects proved too technologically difficult and the funding was too sparse for success. Even the CEV, a program that returns to a capsule concept with a blunt-body ablative heat shield and parachutes (or perhaps a Rogallo wing) to return to Earth (or, perhaps, the ocean), proved a challenge for engineers. The recovery of scientific sample return missions to Earth, both with the loss of Genesis and the successful return of Stardust, suggests that these issues are not exclusive to the human space flight community. As this work is completed, NASA has embarked on the Commercial Crew Development (CCDev) program in which four firms are competing for funding to complete work on their vehicles:

- Blue Origin, Kent, WA—a biconic capsule that could be launched on an Atlas rocket.
- Sierra Nevada Corporation, Louisville, CO—Dream Chaser lifting body, which could be deployed from the Virgin Galactic WhiteKnightTwo carrier aircraft for flight tests.
- Space Exploration Technologies (SpaceX), Hawthorne, CA—Dragon capsule spacecraft; also a partial lifting body concept to be launched on the Falcon 9 heavy lifter.
- The Boeing Company, Houston, TX—a 7-person spacecraft, including both personnel and cargo configurations designed to be launched by several different rockets, and to be reusable up to 10 times.

These new ideas and a broad set of actions stimulated through the CCDev program suggest that reentry and recovery from space remains an unsettled issue in space flight. This book's concluding chapter suggests that our understanding of the longstanding complexities associated with returning to Earth safely has benefited from changes in technology and deeper knowledge of the process; however, these issues are still hotly debated and disagreement remains about how best to accomplish these challenging tasks. Engineers have had success with several different approaches to resolving the challenges of reentry and landing. Discovering the optimal, most elegant solutions requires diligence and creativity.

This history seeks to tell this complex story in a compelling, sophisticated, and technically sound manner for an audience that understands little about the evolution of flight technology. Bits and pieces of this history exist in other publications, but often overlooked is the critical role these concepts played in making a safe return to Earth possible. Moreover, the challenges, mysteries, and outcomes that these programs' members wrestled with offer object lessons in how earlier generations of engineers sought optimal solutions and made tradeoffs. With the CCDev program—a multiphase program intended to stimulate the development of privately operated crew vehicles to low-Earth orbit currently underway—NASA is returning to a capsule concept for space flight. This may prove a significant development, and this history could help enlighten the NASA team about past efforts and the lessons learned from those efforts.

Robert H. Goddard working on one of his rockets in the 1930s. NASA.

CHAPTER 1

Creating Ballistic Reentry

The atmosphere surrounding Earth protects and supports life on the planet, but it also makes space flight more difficult. It is said, only half-jokingly, that getting to orbit is halfway to anywhere because of the considerable amount of energy necessary to go beyond the gravity well of this planet.[1] Generally over-looked, however, is just how difficult it is to return from orbit. All of the energy expended to get to orbit dissipates on the way back to Earth, usually in the form of extreme heating. In addition to the aerodynamic concerns of high-speed flight, there are serious thermodynamic issues with a 17,500-miles-per-hour (mph) plunge through Earth's atmosphere.[2] Unlike the relatively long gesta-tion periods of airplanes and launch vehicles, the technology needed to survive reentry matured quickly, largely in response to national security concerns. The warheads developed for ballistic missiles during the Cold War led directly to the capsules that first allowed humans to venture into space. Of course, it is important to recognize that the requirements for human entry are significantly

---

1. Basic histories of rocketry include Roger D. Launius and Dennis R. Jenkins, eds., *To Reach the High Frontier: A History of U.S. Launch Vehicles* (Lexington, KY: University Press of Kentucky, 2002); David Baker, *The Rocket: The History and Development of Rocket and Missile Technology* (New York: Crown Books, 1978); Frank H. Winter, *Rockets into Space* (Cambridge, MA: Harvard University Press, 1990); Wernher von Braun, Frederick I. Ordway III, and Dave Dooling, *History of Rocketry and Space Travel* (New York: Thomas Y. Crowell Co., 1986 ed.); Eugene M. Emme, ed., *The History of Rocket Technology: Essays on Research, Development, and Utility* (Detroit, MI: Wayne State University Press, 1964); G. Harry Stine, *Halfway to Anywhere: Achieving America's Destiny in Space* (New York: M. Evans and Co., 1996).

2. Representative technical studies of the challenges of reentry and recovery include Patrick Gallais, *Atmospheric Reentry Vehicle Mechanics* (New York: Springer, 2007); Ashish Tewari, *Atmospheric and Space Flight Dynamics: Modeling and Simulation with MATLAB® and Simulink®* (Boston: Birkhäuser Boston, 2009); Frank J. Regan, *Dynamics of Atmospheric Reentry* (Reston, VA: American Institute for Aeronautics and Astronautics [AIAA], 1993); Wilber L. Hankey, *Reentry Aerodynamics* (Reston, VA: AIAA, 1988); and John David Anderson, *Hypersonic and High Temperature Gas Dynamics* (Reston, VA: AIAA, 2006).

more restrictive than those for warheads. Accordingly, the Mercury capsule could not simply be a warhead shell reconfigured for human occupancy. The heat shield required for the longer, lower g entries drove the development of several new technologies. Despite this, the technologies for ballistic reentry paved the way for human space flight and reentry.[3]

After the Columbia accident in 2003, it became popular to ask: Why does the Space Shuttle have wings? The reentry of spacecraft using ablative heat shields and parachutes seemingly worked well, so why abandon it? Many pointed to the presumably superior safety record of the Mercury, Gemini, and Apollo capsules and questioned the apparent change in course of using wings protected from melting by fragile thermal ceramic tiles on the Space Shuttle.[4]

However, it stands to reason that the capsules were the anomaly created because the schedule of the space race did not allow the creation of launch vehicles large enough to loft anything more sophisticated. In contrast, reusable, usually winged space vehicles—often discussed since Robert Goddard, Konstantin Eduardovich Tsiolkovskiy, and Hermann Oberth began writing about space flight during the early part of the 20th century—dominated thinking on the subject until the 1950s and the dawn of the space age. From the beginning, the spaceplane featured prominently in plans, although it should be noted that many (perhaps most) of the early concepts were actually long-range suborbital aircraft and not truly spacecraft. In part because of the dawning realization of the technology involved, realistic visions of true spacefaring vehicles did not begin to appear until the early 1950s.

Early Ideas About Spaceplanes

For many years after scientists and engineers began planning for space flight in the 20th century, the dominant vision of how to achieve this was via orbital spaceplanes that could be launched like rockets or perhaps even like an airplane

---

3. *Reentry Studies*, two vols., Vitro Corporation report no. 2331-25, November 25, 1958, and the National Advisory Committee for Aeronautics (NACA) Conference on High-Speed Aerodynamics, Ames Aeronautical Laboratory, Moffett Field, CA, March 18–20, 1958, A Compilation of the Papers Presented, both NASA technical reports.

4. It is true that no U.S. capsule crew was lost to any cause, although several came close. However, it is impossible to determine if the capsules were really safer than the Shuttle simply because they did not fly often enough to result in any meaningful statistics. Soyuz, another capsule design that has flown almost as often as the Space Shuttle, has a very slightly, statistically irrelevant, worse safety record.

from a runway. Those early space flight engineers envisioned the spaceplanes flying into orbit, undertaking their missions, and returning to land on Earth like airplanes at an airport. These winged orbital vehicles represented an extension of concepts associated with airplanes, including reusability; precise, controlled horizontal landings; acceleration to speeds sufficient to escape Earth's gravity well (approximately 7.5 miles per second); and orbital velocities of 17,500 mph. Some concepts of these vehicles imagined they would be launched vertically like rockets while others speculated they would not be; and some concepts envisioned wings while others described lifting bodies in which the vehicle's shape, rather than wings, would provide lift during reentry and landing.

The objective of airplanelike space operations dominated thinking about space flight from the beginning and had a profound influence on the development of reentry and recovery technology. Indeed, this dominant paradigm for space flight drove many thinkers to advocate reusable winged vehicles until the 1950s, when technological experimentation and the crushing pressure of the Cold War prompted its abandonment in favor of more readily achievable ballistic technologies. Nevertheless, until the recent past, NASA has pursued this concept as its holy grail, jettisoning the ballistic model used successfully in the Mercury, Gemini, and Apollo programs in favor of the reusable Space Shuttle that pointed efforts toward the spaceplane conception so long desired.[5]

At the same time, the goal of all these studies was to achieve (1) pinpoint recovery and (2) reusability, and wings were only one way to achieve these ends. These goals were temporarily set aside during the space race when it became obvious that a capsule shape with one-time use and a water recovery could accomplish the human-in-space goal quicker. However, there is a powerful psychological appeal to incorporating wings on spacecraft because it represents a tried-and-true method of achieving flight demonstrated from the Wright brothers' era to the present. The primary objective of achieving an operational, low-cost space-access system drove this trend toward spaceplanes. But low-cost space flight has not developed as expected, and that may be an underlying reason that the nonreusable, water recovery concepts reemerged with the Constellation effort in the first decade of the 21st century. It is possible that when a true need for high-volume space-access flights arises, the soft-landing, reusable spacecraft concept will again emerge. Will this be achieved with wings

---

5. Roger D. Launius, "Between a Rocket and a Hard Place: The Challenge of Space Access," in W. Henry Lambright, ed., *Space Policy in the 21st Century* (Baltimore, MD: Johns Hopkins University Press, 2002), pp. 15–54.

or landing rockets, with gliding parachutes or rotors, or through a separate approach? Only time will tell.

As a schoolteacher in Moscow, Konstantin Tsiolkovskiy began writing about rockets and space travel in 1898, when he submitted his "Investigating Space with Rocket Devices" article to the Russian *Nauchnoye Obozreniye* (*Science Review*) journal. This article presented years of calculations that laid out many of the principles of modern space flight and opened the door to future writings on the subject. There followed a series of increasingly sophisticated studies on the technical aspects of space flight. In the 1920s and 1930s, Tsiolkovskiy was especially productive, publishing 10 major works, clarifying the nature of bodies in orbit, developing scientific principles behind reaction vehicles, designing orbital space stations, and promoting interplanetary travel. The most important of these works was his *Plan of Space Exploration*, in which he described rocket-propelled airplanes with wings—a forerunner of the Space Shuttle. This led Tsiolkovskiy to deal with the aerodynamics of reentry in an exceptionally cursory manner. He also expanded the scope of studies on many principles commonly used in rockets today, which include specific impulse to gauge engine performance, multistage boosters, fuel mixtures such as liquid hydrogen and liquid oxygen, the problems and possibilities inherent in microgravity, the promise of solar power, and spacesuits for extravehicular activity. Significantly, Tsiolkovskiy never had the resources—perhaps not even the inclination—to experiment with rockets himself. During Tsiolkovskiy's lifetime, much of this work went unnoticed outside Russia.[6]

In the West, one of the first popular spaceplane concepts was Robert Goddard's turbine rocket ship, which appeared in both *Scientific American* and *Popular Science* in 1935. Famous for his successful efforts to build liquid-fueled rockets—the first such launch taking place in 1926, with the succession of progressively more sophisticated rockets following until his death in 1945—Goddard designed his vehicle with elliptical wings and a unique dual-mode propulsion system. When outside the atmosphere, the vehicle used a liquid-fueled rocket engine, as would be expected of a vehicle designed by the person who literally invented the technology. However, when the vehicle was flying within the atmosphere, two turbines moved into the exhaust stream of the rocket engine, driving conventional propellers affixed to either wing via long drive shafts. It was obvious from both the description and the limited artwork that Goddard was not an aircraft designer, and the concept had several major

---

6. Roger D. Launius, *Frontiers of Space Exploration*, 2nd ed. (Westport, CT: Greenwood Press, 2004), pp. 112–113; James T. Andrews, *Red Cosmos: K.E. Tsiolkovskii, Grandfather of Soviet Rocketry* (College Station, TX: Texas A&M University Press, 2009).

flaws. Nevertheless, it helped popularize the idea that space vehicles would be extensions of normal aircraft, complete with wings, tails, and landing gear.[7]

Meanwhile, others were putting form to the theoretical concept of space-planes. The Viennese scientist Max Valier believed that rocket engines would eventually replace the internal combustion engine and lead, by natural evolution, to winged spaceplanes that would travel back and forth to Earth orbit and the planets.[8] In early 1928, Valier convinced Fritz A. H. von Opel, heir to the Adam Opel automobile empire, that the rocket could power a car as a publicity stunt to increase sales. The pair soon moved on to what Valier really wanted to do: build a rocket-powered aircraft. Using two 44-pound-force black-powder rockets attached to a 14-foot-long glider, Friedrich Stamer made the first piloted rocket-powered flight on June 11, 1928. Unfortunately, one of the rockets exploded during the second flight and fire consumed the glider. This did not stop the experiments, however, and on September 30, 1929, von Opel flew a 16-foot-long glider for almost 2 miles in front of a large crowd in Frankfurt am Main, Germany. A hard landing destroyed the glider and injured von Opel. The concept of rocket-powered aircraft became firmly established, if not well demonstrated. Later, during World War II, all combatants developed rocket-power-assisted technologies, usually called jet-assisted at the time, for aircraft to take off from short runways.[9]

At a fundamental level, the desire to build a spaceplane was inextricably linked to the development of hypersonic flight technology, which was in its infancy prior to World War II. Indeed, many viewed even hypersonic flight as

---

7. Robert H. Goddard, "A New Turbine Rocket Plane for the Upper Atmosphere," *Scientific American*, March 1932; *Popular Science*, December 1932.

8. In 1923, Valier was inspired by Hermann Oberth's book *Die Rakete zu den Planetenräumen* (*By Rocket into Interplanetary Space*) to write a similar work to explain the concepts in terms that could be understood by laypersons. With Oberth's assistance, he published *Der Vorstoß in den Weltenraum* (*The Advance into Space*) in 1924. Tragically, Valier died in a laboratory accident on May 17, 1930, when an experimental oxygen-alcohol rocket engine exploded on its test stand.

9. Richard P. Hallion, "In the Beginning was the Dream ...," preface to *The Hypersonic Revolution: Case Studies in the History of Hypersonic Technology, Volume 1, From Max Valier to Project PRIME (1924–1967)* (Bolling Air Force Base [AFB], DC: USAF Histories and Museums Program, 1998), pp. xi–xii; *Max Valier: A Pioneer of Space Travel*, from the German *Max Valier: ein Vorkampher der Weltraumfahrt, 1895–1930*, translated by NASA as TT F-664 (Washington, DC: NASA, 1976), pp. 81–97, 130–135, 248; Walter J. Boyne, "The Rocket Men," *Air Force Magazine* (September 2004): 107–110. This was, apparently, sufficient for von Opel. He retired to Switzerland and lived off the income of the fortune his father received for selling Adam Opel A.G. to General Motors. Opel died on April 8, 1971, more than 40 years after his only rocket flight.

difficult and impractical, realizing as they did that the characteristics exhibited by objects moving at Mach 5 or greater were exceptionally, perhaps even unfathomably, complex. They knew that when an aircraft reached hypersonic speed, the amount of pressure it generated increased to 25 times the atmospheric pressure.[10]

Researchers also found that as shock waves at those speeds grew in intensity, their angles bent as they passed between the shock wave and the surface of the vehicle. This region of multiple shock waves, called the shock layer because it was very small, could interact with the boundary layer. The body of the aircraft then experienced significant turbulence, and the air around the aircraft became a swarming jumble of hot gases whose intense heat transferred to the aircraft. Once the shock waves have conformed to the windward shape of the vehicle, however, the performance and stability are well described by simple Newtonian flow theory. Even though stability and controllability were easily achievable, heat transfer remained complex and difficult to characterize and control.[11]

The interaction between the shock layer and the boundary layer affected drag and caused unusual thermal reactions, contributing significantly to aerodynamic heating, a product of the friction of the high-speed airflow over the vehicle's surface. Because of these complex interactions, the behavior and characteristics of hypersonic airflow proved extremely difficult to analyze. Only Goddard, Sänger, von Braun, and a handful of other space flight advocates and experimenters worked seriously to understand hypersonic flight and, mostly, to learn how to achieve the speeds necessary to reach Earth's orbit.[12]

None of this had much to do with the reality of getting to orbit. It was, however, critical to the process of reaching into space; without achieving that goal, reaching reentry and recovery from space flight was a moot issue.

---

10. Dennis R. Jenkins, *Hypersonics Before the Shuttle: A Concise History of the X-15 Research Airplane* (Washington, DC: NASA SP-2000-4518, 2000), pp. 7–8.

11. The technical aspects of hypersonic aerodynamics may be found in Theodore A. Talay, *Introduction to the Aerodynamics of Flight* (Washington, DC: NASA SP-367, 1975), chapter 7.

12. For an overview see Richard P. Hallion, ed., *The Hypersonic Revolution: Eight Case Studies in the History of Hypersonic Technology*, 3 vols. (Washington, DC: Air Force History and Museums Program, 1998). This work contains nine studies of hypersonic research and development (R&D) programs: the X-15, the X-20A Dyna-Soar, winged reentry vehicles, ASSET, Project PRIME, the Scramjet, lifting bodies, the Space Shuttle, and the National Aero-Space Plane. On larger issues in the history of hypersonic technology, see Curtis Peebles, "The Origins of the U.S. Space Shuttle–1," *Spaceflight* 21 (November 1979): 435–442; "The Origins of the U.S. Space Shuttle–2," *Spaceflight* 21 (December 1979): 487–492.

Sänger, Bredt, and the Silverbird

Perhaps Sänger made the most interesting of the early studies into orbital winged space flight. Inspired by Oberth's book *By Rocket into Planetary Space*, Sänger changed his university studies from civil engineering to aeronautics. While he was a doctoral candidate at the Viennese Polytechnic Institute in 1929, Sänger conceptualized a reusable, rocket-powered spaceplane with straight wings. However, like many of the other early spaceplane concepts, most of the Sänger designs were suborbital aircraft, not true spacecraft. In late 1933, Sänger privately published *Techniques of Rocket Flight*, which contained a detailed description of the vehicle that was called the "Silverbird."[13] In collaboration with mathematician Dr. Irene Bredt, whom he later married, Sänger continued to refine this design for the next 30 years.[14]

Propelled by a liquid-fueled rocket engine, the Silverbird was capable of reaching 6,600 mph at an altitude of 100 miles. Due to the relatively low speed, this was not an orbital vehicle, despite the high altitude it theoretically could achieve.[15] Instead, the Silverbird used a suborbital trajectory to deliver a payload halfway around the world using a technique that included a series of semiballistic skips off the atmosphere that Sänger called "dynamic soaring." As the vehicle followed a ballistic path from space into the upper atmosphere, each skip reduced the forward speed slightly, and, consequently, each skip was a little lower and shorter. Sänger believed this strategy would increase range

---

13. Interestingly, Sänger used the same publisher as Valier, and it took 4 years to pay off the resulting printing charges.

14. Eugen Sänger and Irene Bredt, "The Silverbird Story: A Memoir," in R. Cargill Hall, ed., *Essays of the History of Rocketry and Astronautics: Proceedings of the Third Through Sixth History Symposia of the International Academy of Astronautics*, vol. 1 (Washington, DC: NASA, 1977), pp. 195–288; Eugen Sänger, *Rocket Flight Engineering from the 1933 German Raketenflugtechnik*, translated by NASA as TT-F-223 (Washington, DC: NASA, 1965); Willey Ley, *Rockets, Missiles, and Space Travel* (New York: The Viking Press, 1957), pp. 429–434.

15. Eugen Sänger, *Recent Results in Rocket Flight Technique*, translated by the NACA from a 1934 German paper as TM-1012 (Washington, DC: NACA, 1942). Reports on the payload of the Silverbird vary widely, even within Sänger's papers. Some estimates are as low as 660 pounds for the antipodal version, and others are as high as 16,000 pounds. It is difficult to tell what is realistic, but both of these extremes are unlikely.

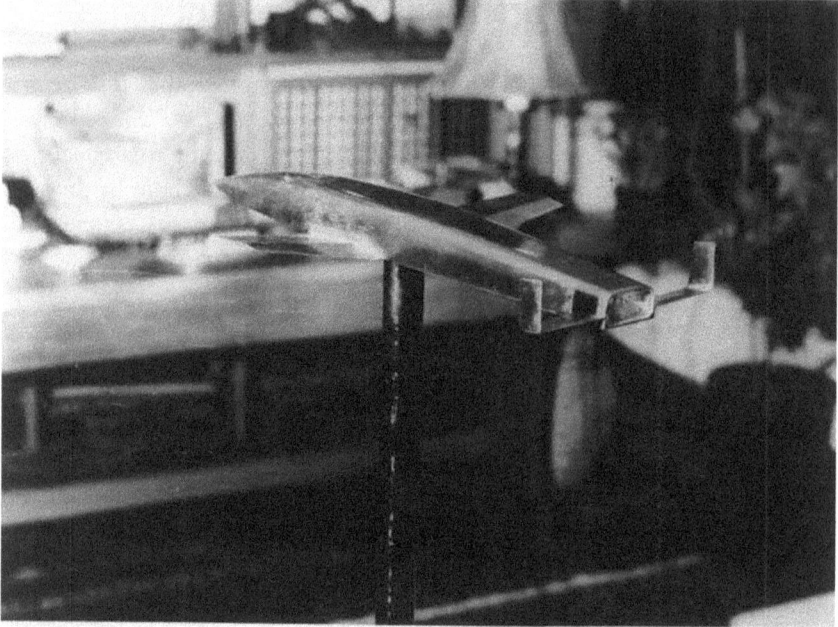

The Sänger Silverbird in wind tunnel testing in the United States after World War II. NASA.

and significantly reduce the reentry[16] thermal load by allowing heat to radiate into space after each skip.[17] Despite this hypothesis, Sänger did little actual investigation into reentry conditions, concentrating more on the propulsive aspects of the vehicle. Of course, even if he had wanted to study reentry effects, little theoretical and no practical basis existed to do so.[18]

Accordingly, while there was some understanding of pressures and heating in the period before World War II, the majority of studies into the challenges of hypersonic flight had to wait until the postwar era. For one, the large V-2

---

16. The terms "entry" and "reentry" are often used interchangeably but, in fact, refer to different maneuvers. An entry is made following an orbital or superorbital flight; a reentry is the concluding maneuver of a suborbital (or nonorbital) flight. Still, for the purposes of this monograph, the terms possess no particular significance other than the velocities at which they are accomplished.

17. Hence the term "antipodal glider," which is frequently applied to the Sänger designs. *Webster's* defines antipodal as "a point on the opposite side of the Earth or moon...."

18. Eugen Sänger and Irene Bredt, *A Rocket Drive for Long-Range Bombers*, from the German *Über einen Raketenantrieb für Fernbomber*, translated by the Naval Technical Information Brach, Bureau of Aeronautics as CGD-32 (Washington, DC: U.S. Navy, 1952).

rockets developed in Germany during World War II forced the development of concept studies with promising results.[19]

Around the same time, in 1946, the work of German engineers Sänger and Bredt reached the U.S. Navy's Bureau of Aeronautics, changing many people's perspectives about the possibility of hypersonic flight. The two engineers argued that a rocket-powered hypersonic aircraft could be built with only minor advances in technology. NACA engineers found this concept quite stimulating, and along with studies by German and Russian researchers, this prompted engineers such as John Stack and John V. Becker at the Langley Memorial Aeronautical Laboratory to begin the lengthy process of gaining approvals to explore this flight regime.[20]

Skip-gliding remained in favor for nearly 20 years, until research at the NACA Ames Aeronautical Laboratory in the 1950s showed the technique would not reduce heating as effectively as would a direct reentry. While radiating heat into space during skip-gliding proved a highly effective and useful method of rejecting heat and has subsequently proven a primary method of controlling temperatures in the true space environment, the direct reentry approach proved more expeditious for those early efforts. Although peak temperatures during skip-gliding would be slightly lower than those during direct reentry, the time the vehicle spent exposed to those temperatures was considerably longer (cumulatively, through all the skips). This proved a problem for heat-sink and ablative thermal protection systems, but research soon devised hot structures that used radiation or active cooling where the temperatures were stabilized and not subjected to total heat-load restrictions.

The First Ballistic Missiles

During the 1920s and 1930s, organizations of rocket enthusiasts emerged in several European countries, with the most influential being the German Society for Space Travel, or Verein für Raumschiffahrt (VfR). Spurred by the theoretical arguments of Oberth and Valier, the VfR emerged soon after its founding,

---

19. The standard work on this subject is Michael J. Neufeld, *The Rocket and the Reich: Peenemünde and the Coming of the Ballistic Missile Era* (New York: The Free Press, 1995).

20. On Sänger, see Irene Sänger-Bredt, "The Silverbird Story," *Spaceflight* 15 (May 1973): 166–181. Their pathbreaking study of hypersonic flight, translated from German and made available to the NACA researchers, is Sänger, Bredt, "*Über einen Raketenantrieb für Fernbomber*" (Deutsche Luftfahrtforschung U93538, 1944), translation CGD-32, Technical Information Branch, BuAer, Navy Department.

The German Army in World War II preparing the V-2 for a test. NASA.

on July 5, 1927, as the world's leading space-vehicle design group. Specifically organized to raise money, it tested Oberth's rocketry ideas. It soon proved successful in building a base of support in Germany, publishing a magazine and scholarly studies, and in constructing and launching small rockets. From the beginning, one of the VfR's strengths, however, was its ability to publicize both its activities and the dream of space flight.[21]

The VfR made good on some of those dreams on February 21, 1931, when it launched the liquid-oxygen/methane-liquid-fueled rocket HW-1 near Dessau, Germany, to an altitude of approximately 2,000 feet. The organization's public relations arm went into high gear after this mission and emphasized the launch's importance as the first successful European liquid-fueled rocket flight.[22] Von Braun, then a neophyte learning the principles of rocketry from Oberth and Valier, was both enthralled with this flight and impressed with the publicity it engendered. Later, he became the quintessential and movingly eloquent advocate for the dream of space flight and a leading architect of its technical development. He began developing both skills while working with the VfR.[23] In the end, the VfR conducted numerous static firings of rocket engines and launched several small rockets.

As Germany rearmed in the 1930s, its army collected a host of rocket engineers, many from the VfR, at a military test facility at Peenemünde, Germany, on the Baltic coast, and initiated the development of a short-range ballistic missile under the tutelage of von Braun.[24] By 1943, after numerous frustrations, they had a working vehicle. Joseph Goebbels christened the rocket (called Assembly 4, or A-4, at Peenemünde) "Vergeltungswaffe Zwei" (Vengeance Weapon Two), or V-2. The missile became operational in September 1944 and over 3,500 were launched in aggression, mainly against Antwerp and London.

---

21. The standard work on the rocket societies is Frank H. Winter, *Prelude to the Space Age: The Rocket Societies, 1924–1940* (Washington, DC: Smithsonian Institution Press, 1983). A briefer discussion is available in Frank H. Winter, *Rockets into Space*, pp. 34–42.

22. Winter, *Rockets into Space*, p. 37.

23. Wernher von Braun, "German Rocketry," in Arthur C. Clarke, ed., *The Coming of the Space Age* (New York: Meredith Press, 1967), pp. 33–55. Von Braun's public relations skills were exceptional throughout his career. Evidence of this can be found in the more than 8 linear feet of materials by von Braun held in the Biographical Files of the NASA Historical Reference Collection, NASA History Program Office, NASA Headquarters, Washington, DC.

24. By the accepted definitions used in the industry, a missile is a guided vehicle while a rocket is an unguided vehicle. The guidance can take the form of a human pilot or a mechanical or electronic autopilot.

The designers of the first modern ballistic missile did not have much need to worry about reentry effects. The A-4 reached an altitude of only 50 to 60 miles, after which it descended on a ballistic trajectory with a maximum velocity of approximately Mach 4 and a range of 200 miles. Although this velocity made the weapon invulnerable to anti-aircraft guns and fighters, the temperatures it experienced while streaking through the thick atmosphere near Earth's surface were well within the heat and strength capability of conventional steel and aluminum alloys.[25]

While deployed too late to alter the outcome of the war, the A-4 not only radically changed the concept of weapon delivery but also offered the promise of space flight in the not too distant future. It became the basis for all that followed in the Soviet Union and United States. By the end of World War II, the popular conception of a missile along the lines of the German A-4—a slender cylinder with a pointed nose and sweptback fins—had become the norm. This basic shape, used for decades in various books and movies, followed the general streamlining trends in aviation that began in the 1920s. The experience with the A-4 had shown the shape worked adequately, although few truly appreciated the relatively low velocities achieved by the missile.

During the late 1940s, the United States began developing an increasingly sophisticated family of ballistic missiles. As a direct result of war booty captured from the Germans at the end of World War II, the United States used the V-2 first stage with a WAC Corporal rocket as a second stage and tested hypersonic concepts at the White Sands Proving Ground, NM. The V-2/WAC Corporal combination became the first manufactured object to achieve hypersonic flight. On February 24, 1949, its upper stage reached a maximum velocity of 5,150 mph—more than five times the speed of sound—and a 244-mile altitude. The vehicle, however, burned up on reentry, and only charred remnants remained, demonstrating the need to explore the challenges of reentry and recovery from space.[26]

First launched in 1953, the Redstone was a direct descendant of the German A-4, designed largely by the same team from Peenemünde led by von Braun. The missile flew about as high and far as the A-4, but it carried a much larger payload (6,500 versus 2,500 pounds) and reached a maximum velocity of Mach

25. Walter R. Dornberger, V-2 (New York: The Viking Press, 1954), p. 245; Gregory P. Kennedy, *Vengeance Weapon 2: The V-2 Guided Missile* (Washington, DC: Smithsonian Institution Press, 1983).

26. David H. DeVorkin, *Science with a Vengeance: How the Military Created the US Space Sciences After World War II* (New York: Springer-Verlag, 1992), pp. 45–58; John D. Anderson, Jr., *A History of Aerodynamics, and Its Impact on Flying Machines* (Cambridge, England: Cambridge University Press, 1997), pp. 437–439.

Bumper 8, on July 24, 1950, was the first launch from Cape Canaveral, FL. NASA.

5.5 during reentry. Improved versions, called Jupiter, that were developed a few years later extended the maximum range to 1,500 miles and reached Mach 15 during reentry from a 390-mile peak altitude. This significant increase in performance brought new problems for the engineers as they explored how to recover portions of the vehicle.

The Problem of Reentry

Given the technology available during the 1950s—indeed, even 60 years later—getting into space has been largely an exercise in brute force. Sufficient power, usually in the form of rocket engines, must be available to accelerate the vehicle to slightly over 17,500 mph to attain a stable, low-Earth orbit. Unlike aircraft that can match their rate of climb to the available thrust, conventional rockets must always accelerate at greater than 1 g since they rely solely on power, not lift, to reach in space. The only exception to this 1 g–plus requirement for vertical liftoff is a winged rocket (like the X-15) that can fly trajectories that are not vertical and use lift as well as thrust. The exit trajectory is usually very steep, so the vehicle is well above the dense atmosphere before it reaches high speeds, a tactic that minimizes both drag and aerodynamic heating.

Oddly, reentry went largely unnoticed in most early literature on space flight. Although many people had speculated on how to get into space, relatively few were thinking about how to get back. One exception was Walter Hohmann, a German civil engineer and member of the VfR, who in 1925 wrote *Die Erreichbarkeit der Himmelskörper* (*The Attainability of Celestial Bodies*).[27] Although the book was primarily concerned with the derivation of optimum transfer trajectories from Earth to other planets, Hohmann also thought about how to return to Earth, and he was especially concerned about the effects of atmospheric heating during reentry.[28] Unfortunately, Hohmann lacked any practical knowledge of and any meaningful way to investigate the environment in which a vehicle would have to operate in returning from space. His work was entirely theoretical. Although he failed to come up with specific solutions to the problem, Hohmann categorized the problem and theorized several techniques that returning spacecraft might use, including variable-geometry wings and external insulation. Despite the work by Hohmann and a few others, none of the technologies available at the time would have permitted building a reentry vehicle, even if a means of launching it had existed.[29]

Twenty years later, at the end of World War II, it was apparent that space flight was not too far off, although it shocked nearly everyone when it arrived as soon as it did. One of the first serious postwar studies of space flight was the *Preliminary Design of an Experimental World-Circling Spaceship* by Project

---

27. Walter Hohmann, *Die Erreichbarkeit der Himmelskörper* (*The Attainability of Celestial Bodies*), (Munich, Bavaria: R. Oldenbourg, 1925), translated by NASA (Washington, DC: NASA TT F-44, 1960); William I. McLaughlin, "Walter Hohmann's Roads in Space," *Journal of Space Mission Architecture* issue 2, Jet Propulsion Laboratory (JPL), (fall 2000): 1–14.

28. Despite working largely with unknowns, in 1925, Hohmann managed to define the most economical path (although not necessarily the shortest or fastest) for a spacecraft to take from one planet to another. Known today as a "Hohmann Transfer Orbit," Hohmann showed that elliptical orbits tangent to the orbits of both the departure and target planets require the least energy.

29. Hallion, "The Path to the Space Shuttle: The Evolution of Lifting-Reentry Technology," unpublished manuscript in the files of the Air Force Flight Test Center (AFFTC) History Office (April 1983).

RAND at Douglas Aircraft. As with Sänger and Hohmann, RAND attempted to explore the entry environment but lacked any real basis for developing a theory. Nevertheless, the study concluded the following:

> An investigation was made of the possibility of safely landing the vehicle without allowing it to enter the atmosphere at such great speeds that it would be destroyed by the heat of air resistance. It was found that by the use of wings on the small final vehicle, the rate of descent could be controlled so that the heat would be dissipated by radiation at temperatures the structure could safely withstand. The same wings could be used to land the vehicle on the surface of the earth.[30]

It was a credible endorsement of a spaceplane. Of course, it was not that simple. Although the RAND engineers expected to be able to control the rate of descent by using the wings and radiating the heat while still at high velocity, this approach ignored the characteristics of the atmosphere at extreme altitudes, which were largely unknown at the time. This is the same concept that was used on the X-20 Dyna-Soar glider in the late 1950s and early 1960s and was successfully demonstrated on the ASSET vehicle.[31]

Ultimately, most engineers working on the problems of space flight were convinced that it would prove impossible to fly back from orbit, at least in the short run.[32] Even the legendary Theodore von Kármán was worried. In his 1954 history of aeronautics, von Kármán observed of a reentry vehicle, "At such speeds, probably even in the thinnest of air, the surface would be heated beyond the temperature endurable by any known material. This problem of the temperature barrier is much more formidable than the problem of the sonic barrier."[33] Thus was born the thermal barrier terminology that would define much of the aeronautical research during the late 1950s and early 1960s.

30. F.H. Clauser, "Preliminary Design of an Experimental World-Circling Spaceship," Douglas Aircraft Company, Santa Monica Engineering Division, RAND Report No. 1 (2 May 1946), pp. VII and 192–195.

31. Ibid., p. 198.

32. Ibid., pp. 221, 225–229.

33. Theodore von Kármán, *Aerodynamics: Selected Topics in the Light of their Historical Development* (Ithaca, NY: Cornell University Press, 1954), p. 189. This excellent work was republished by Dover Publications in 2004.

Nuclear Warheads and the Blunt-Body Theory

Initially, all objects returned from very high altitudes using a ballistic reentry in which the force was parallel to the line of flight and the trajectory was always in the form of a parabola. Arrows and artillery projectiles have long used ballistic trajectories, and long-range missiles initially followed suit. The peak altitude of the early missiles, such as the A-4 and the Redstone, was low enough that the vehicle did not reach velocities that produced atmospheric friction sufficient to create a serious heating problem. However, as the maximum range of missiles grew longer, the peak altitudes and velocities increased and heating became a major concern. To attain a range of one-quarter the circumference of Earth, an object must reach a velocity of about 23,000 feet per second (ft/sec) (15,500 mph). The vehicle then has about eight times the kinetic energy required to turn ice into steam.[34] During reentry, almost all of this kinetic energy converts into heat through atmospheric heating.[35]

The steep trajectories used by the initial medium-range ballistic missiles resulted in a reentry in which temperatures exceeded 12,000 degrees Fahrenheit (°F) within the stagnation zone immediately in front of the vehicle. This is twice as hot as the surface of the Sun. The heat generated outside the boundary layer by shock wave compression, which was not in contact with the vehicle, dissipated harmlessly into the surrounding air. The heat within the boundary layer, and in direct contact with the structure, however, was hot enough to melt the vehicle.

At Convair, Karel Bossart and his team of engineers were developing the Atlas intercontinental ballistic missile (ICBM), the first of the long-range ballistic missiles. In a radical departure from previous missiles, Bossart decided that only the warhead would return to Earth, not the entire missile. Convair knew the warhead—usually referred to as a "reentry vehicle" since it sounded less sinister—would need protection from the thermal effects of reentry, but it had little test-validated theory to allow a design to proceed. Using a digital computer, one of the first applications of the device as a design simulation tool, Convair engineers began examining various shapes for the warhead.

Since the 1920s, airplanes and the missiles that followed became increasingly streamlined. The goal, of course, was to go faster by creating less drag. These lessons seemed particularly applicable to the fastest vehicle yet: the

---

34. It takes about 1,800 BTU to convert one pound of ice into 1,000 °F steam.

35. H. Julian Allen, "The Aerodynamic Heating of Atmosphere Entry Vehicles," a paper prepared for the Symposium on Fundamental Phenomena in Hypersonic Flow at Cornell Aeronautical Laboratory, Buffalo, NY, June 25–26, 1964.

ballistic missile. When Convair engineers fed data into their digital computer, they did so with the bias that more streamlined was better. So, unsurprisingly, the computer-generated results favored a slender, pointed-nose reentry vehicle that looked much like the Bell X-1 that first broke the sound barrier.

Convair and NACA researchers then tested the shape in the few hypersonic wind tunnels that existed and used rocket-boosted free-flight models at the NACA Pilotless Aircraft Research Station on Wallops Island, VA. These tests produced an unexpected result: the warhead would absorb so much heat that it would vaporize as it reentered the atmosphere. Suddenly, the earlier experiences with the medium-range warheads began to make sense.

This realization, that the slender aircraft body suited to supersonic flight was inappropriate for the hypersonic flight of a ballistic missile, led weapons designers to explore the phenomenon more thoroughly. This yielded the observation that a blunt-nose body experienced much less heating than a pointed body, which would burn up before reaching Earth's surface. One of the members of a panel charged with the reentry problem was NACA researcher H. Julian Allen.[36] The 42-year-old Allen joined the NACA in 1935 and had been chief of the high-speed research division at NACA's Ames Aeronautical Laboratory, now Ames Research Center, since 1945. Allen enlisted the assistance of Alfred J. Eggers, Jr., a 30-year-old aerodynamicist who had joined the NACA directly out of school in 1944.

The blunt reentry body theory discovered in 1951 by Allen created a stronger shock wave at the nose of the vehicle and dumped a good deal of the reentry heat into the airflow, making less heat available to heat the reentry vehicle itself. This finding was so significant and in such sharp contrast with intuitive thinking that Allen's work fundamentally reshaped the course of reentry and recovery studies and provided the basis for all successful reentry vehicles since.[37]

By June 1952, Allen and Eggers had found a theoretical solution to the aerodynamic heating problems of ballistic reentry vehicles. The two researchers deduced that about half the heat generated by aerodynamic friction transferred into the warhead, quickly exceeding its structural limits. The obvious solution was to deflect the heat away from the vehicle. The breakthrough was in how to accomplish this. In place of the traditional sleek missile with a sharply pointed nose, the researchers proposed a blunt shape with a rounded bottom. When reentering the atmosphere, the

---

36. As told by John Becker, Allen's given name was "Harry," but he disliked the name and always used H. Julian instead. Occasionally, he used Harvey as a nickname, leading to the use of that name in many publications.

37. Glenn E. Bugos, *Atmosphere of Freedom: Sixty Years at the NASA Ames Research Center* (Washington, DC: NASA SP-2000-4314, 2000), pp. 28–30, 46–48.

H. Julian Allen stands beside the observation window of the 8-by-7-foot test section of the NACA Ames Unitary Plan Wind Tunnel. He is best known for his "Blunt Body Theory," which revolutionized the design of ballistic missile reentry shapes. NASA GPN-2000-001778.

rounded body creates a powerful detached shock wave that deflects the airflow and its associated heat outward and away from the reentry vehicle. As Allen and Eggers observed, not only should pointed bodies be avoided, but the rounded nose should have as large a radius as possible. The blunt-body theory was born.[38]

The NACA researchers briefed members of the Atlas ballistic missile development team and select others in September 1952. A secret NACA research memorandum was published for those with appropriate clearances on April 28, 1953, but it would be 5 years before the concept was revealed to the community at large.[39] Interestingly, Allen and Eggers had assumed for the purposes of their research that any warhead would probably use a liquid cooling system to protect it from the residual heat, although the exact nature of an operational thermal protection system was not investigated. The researchers also pointed out that the blunt-body theory worked best for lightweight reentry vehicles and that heavier vehicles likely would need to return, at least partially, to the traditional long, slender shapes once adequate thermal protection systems had been developed.[40]

---

38. H. Julian Allen and Alfred J. Eggers, Jr., "A Study of the Motion and Aerodynamic Heating of Ballistic Missiles Entering the Earth's Atmosphere at High Supersonic Speeds," NACA confidential report RM A53D28, April 28, 1953 (this report was subsequently updated as TN-4047 in 1957 and Report 1381 in 1958); Edwin P. Hartmann, *Adventures in Research: A History of the Ames Research Center, 1940–1965* (Washington, DC: NASA SP-4302, 1972), pp. 216–218. At the time, there was some dispute about exactly who had devised the blunt-body concept. H.H. Nininger, the director of the American Meteorite Museum at Sedona, AZ, claimed he first proposed the blunt nose for reentry vehicles in August 1952. Nininger, a recognized authority on meteorites, based his conclusion on studies of tektites and meteorites. He contended that the melting process experienced by meteorites during their descent through the atmosphere furnished a lubricant that protected them from aerodynamic friction. This letter evidently came to Ames some weeks after Allen and Eggers had completed their study. Despite the contention of Nininger, what Allen wanted to do was exactly the reverse: deliberately shape a reentry body bluntly in order to *increase* air resistance and dissipate a greater amount of the heat produced by the object into the atmosphere. See various letters in the file for H.H. Nininger 1935–1957, NASA History Program Office, Washington, DC.

39. Lloyd S. Swenson, Jr., James M. Greenwood, and Charles C. Alexander, *This New Ocean: A History of Project Mercury* (Washington, DC: NASA SP-4201, 1966), pp. 60–61.

40. H. Julian Allen and Alfred J. Eggers, Jr., "A Study of the Motion and Aerodynamic Heating of Ballistic Missiles Entering the Earth's Atmosphere at High Supersonic Speeds," pp. 12–13. This is precisely what happened. The original Mk II warhead for Atlas was an extreme example of the blunt-body theory, as were the piloted space capsules of Mercury, Gemini, and Apollo. Later warheads, including those in use today, have returned to the slender, pointy shapes originally predicted by Karel Bossart's digital computer, protected by ablative products undreamed of in the early 1950s.

A nonlifting reentry vehicle is measured by its ballistic coefficient (usually designated beta and signified with the Greek symbol "β"), a function of mass, size, and drag coefficient.[41] Vehicles with a high ballistic coefficient—usually long, slender, and with a pointed nose—plunge through the upper atmosphere and experience most of their deceleration in the thick lower atmosphere, where they get very hot for a brief period. Vehicles with a low ballistic coefficient—blunt bodies—experience most of their deceleration in the thin upper atmosphere. They take longer to slow down and generate less heat but experience this heat over a longer period of time. The total heat load (temperature multiplied by time) is essentially the same for both types of vehicles, but the longer, lower-magnitude temperatures of the blunt body were easier for the material available at the time to absorb.

Although the NACA showed that a blunt body was desirable, exactly what shape this should assume was the subject of vigorous debate. Researchers tested spheres, cylinders, and blunted ogives. Allen pioneered much of the practical research on the blunt-body theory and conducted it in an innovative free-flight tunnel at Ames. The $200,000 tunnel had an 18-foot-long test section that was 1 foot wide and 2 feet high. Air was forced into the test section from one direction, and a small model was shot from a compressed-air cannon in the opposite direction. Seven Schlieren cameras provided shadowgraphs that showed the airflow characteristics and shock waves around the model as it passed through the test section. Using this device, researchers could simulate speeds of about Mach 15, which was considerably greater than the few hypersonic wind tunnels or any of the experimental rocket-powered research airplanes.[42]

Eggers contributed another test facility, which was called the Atmospheric Entry Simulator. This was a straight, trumpet-shaped supersonic nozzle 20 inches in diameter and 20 feet long. A hypervelocity gas gun launched a small-scale model upstream through the nozzle into a settling chamber. While in free flight through the contracting nozzle, the model passed through ever-denser air, closely approximating the plunge of a warhead through the atmosphere. Using a model only 0.36 inch in diameter and weighing only 0.005 pounds, Eggers could simulate the aerodynamic heating of an object 3 feet in diameter and weighing 5,000 pounds.[43]

---

41. The ballistic coefficient is derived from the mass of the object divided by the diameter squared that it presents to the airflow divided by a dimensionless constant "I" that relates to the aerodynamics of its shape. The ballistic coefficient has units of pounds per square inch.

42. Alvin Seiff, "A Free-Flight Wind Tunnel for Aerodynamic Testing at Hypersonic Speeds," NACA TR-1222 (May 11, 1955).

43. Swenson, Greenwood, and Alexander, *This New Ocean*, p. 66.

In addition, testing was conducted using the 11-inch hypersonic wind tunnel developed by Becker at Langley and free-flight models at the Pilotless Aircraft Research Station on Wallops Island. The U.S. Air Force (USAF) and Navy also procured 26 three-stage solid-propellant X-17 research vehicles from Lockheed to study reentry techniques up to Mach 15 and 500,000 feet in altitude. The first launch was on April 17, 1956, from Cape Canaveral, FL.[44]

With this basic knowledge, the decision to pursue more research into the hypersonic arena came on June 24, 1952, when the NACA Committee on Aerodynamics passed a resolution to "increase its program dealing with the problems of unmanned and manned flight in the upper stratosphere at altitudes between 12 and 50 miles, and at Mach numbers between 4 and 10." The NACA Executive Committee ratified this decision the following month and appointed a study group led by Clinton E. Brown to ascertain the feasibility of the project.[45]

The report prepared by Brown's committee served as a catalyst for a discussion of hypersonics at the October 1953 meeting of the U.S. Air Force's Scientific Advisory Board (SAB) Aircraft Panel. This panel also provided additional support for hypersonic research. Interestingly, the SAB panel member from Langley, Robert R. Gilruth, director of the Pilotless Aircraft Research Division and later the director of human space flight for NASA, became enthralled with these possibilities and played an important role in negotiating a joint project by the Air Force and the NACA.[46]

As Langley engineer Becker remembered the story in 1968:

> By 1954, we had reached a definite conclusion: the exciting potentialities of these rocket-boosted aircraft could not be realized without major advances in technology in all areas of aircraft design. In particular, the unprecedented problems of aerodynamic heating and high-temperature structures appeared to be so formidable that they were viewed as "barriers" to hypersonic flight.... But in 1954 nearly everyone believed intuitively in the continuing rapid increase in flight speeds of aeronautical vehicles. The powerful new propulsion systems needed for aircraft flight beyond Mach

44. Jay Miller, *The X-Planes: X-1 to X-45* (Hinckley, U.K.: Midland Counties Publishing, 2001), pp. 213–217.

45. John V. Becker, "The X-15 Program in Retrospect," 3rd Eugen Sänger Memorial Lecture, presented at the 1st Annual Meeting, Deutsche Gesellschaft für Luft- und Raumfahrt, Bonn, Germany, December 4–5, 1968, pp. 1–3, copy in the NASA Historical Reference Collection.

46. Jenkins, *Hypersonics Before the Shuttle*, pp. 7–8.

3 were identifiable in the large rocket engines being developed in the long-range missile programs. There was virtually unanimous support for hypersonic technology development.[47]

Therefore, despite technological challenges, the successful hypersonic research led inevitably to greater understanding of the technology needed for safe reentry.

Even though a desirable shape had been determined, the blunt-body theory was, in itself, not a deployable concept. Although it was a major breakthrough, a blunt body still experienced peak temperatures of over 5,000 °F. In the 1950s, the aviation community was using mostly conventional metal alloys, such as Monel K (for the Douglas X-2) and Inconel X (for the North American X-15), but these still could withstand at most 1,200 °F. Allen's assumption that an adequate implementation of a thermal protection system could be designed did indeed place a great enough burden on other researchers and engineers, especially materials scientists, to make it a reality.

Fortunately, the medium-range ballistic missiles experienced such a short-duration reentry that the total heat load could be absorbed by large heat sinks. These heat-sink materials needed to possess certain characteristics. For instance, they needed to be amenable to fabrication into the required shapes. Although, given the small number of warheads being deployed, they did not necessarily have to be mass producible or inexpensive. Most importantly, they had to exhibit a high strength-to-weight ratio at elevated temperatures, have a high melting or sublimation point,[48] have high thermal conductivity and ductility,[49] have a low coefficient of thermal expansion, and be resistant to oxidation. Accordingly, researchers at any number of Government installations and their contractors began investigating heat-sink materials.[50]

Those researchers arranged possible materials into several groups according to shared characteristics and began systematically testing the limits of each group. The first group included heavy ductile metals with high thermal conductivity, such as copper, gold, and silver. These materials have relatively low melting points but maintain their structural integrity and shape under severe heat loads, and they are moldable to almost any shape using conventional machine tools.

---

47. Becker, "The X-15 Program in Retrospect," p. 2.
48. Sublimation is the process by which solids are transformed directly to the vapor state without passing through the liquid phase.
49. A material's ability to withstand force by changing form before fracturing or breaking.
50. Jackson R. Stalder, "The Useful Heat Capacity of Several Materials for Ballistic Nose-Cone Construction," NACA TN-4141 (August 9, 1957), pp. 1–2.

X-15-3 (56-6672) flies over the Mojave Desert in the 1960s. Ship #3 made 65 flights during the program, attaining a top speed of Mach 5.65 and a maximum altitude of 354,200 feet. NASA E-USAF-X-15.

The second group was composed of medium-density refractory metals such as nickel-chromium-iron stainless steels (such as Inconel) and various cobalt alloys developed for jet-engine turbine blades. These metals retain strength at elevated temperatures, have a relatively high resistance to oxidation, and possess low values of specific heat and thermal conductivity. The third group included lightweight metals such as beryllium, with high strength-to-weight ratios, high specific heat and thermal conductivity, high resistance to short-time oxidation, and reasonably high melting points. However, these metals suffered from poor ductility and were difficult to manufacture into complex shapes. The last group included semimetals such as carbon (graphite) that have very high sublimation points, high thermal conductivity and specific heat, and low density. Most of these materials, however, exhibit poor high-temperature oxidation resistance, low structural strength, and are difficult to fabricate.[51]

NACA researchers concentrated on a single member of each group, choosing copper, Inconel X,[52] beryllium, and graphite because they were commercially

---

51. Ibid., pp. 5–6.

52. Inconel X® is a temperature-resistant alloy whose name is a registered trademark of Huntington Alloy Products Division, International Nickel Company, Huntington, WV. It is, for all intents, an exotic stainless steel. Inconel X is 72.5 percent nickel, 15 percent chromium, and 1 percent columbium, with iron making up most of the balance. Although used in a wide variety of applications, it is best known as the primary structural material for the North American X-15 research airplane.

available and relatively well understood. Researchers manufactured a 1-inch-thick slab of each material for a series of tests, and the results were telling. For instance, the front face of each slab was heated to 2,000 °F and the temperature of the rear face was measured to determine the heat absorption capacity. The copper slab had a rear-face temperature of only 190 °F, indicating that the material absorbed most of the thermal load. On the other hand, Inconel X had a rear-face temperature of 1,600 °F, meaning the metal was ineffective at absorbing heat.[53] Beryllium and graphite had rear-face temperatures of 840 °F and 1,240 °F, respectively.[54]

Researchers also tested the effect of slab thickness on the maximum surface temperature during reentry. Inconel X was excluded from this test since the surface of the slab reached its melting temperature regardless of the thickness. The amount of heat absorbed per unit weight of material was measured at a temperature corresponding to 75 percent of the melting (or sublimation) point. In this test, copper absorbed 84 British thermal units (BTU) per pound, beryllium 534 BTU, and graphite 1,980 BTU. This made beryllium and graphite greatly superior to copper as heat-sink materials, especially given that copper was 24 times heavier than graphite and 6 times heavier than beryllium.[55] Given the low performance of the early missiles, weight was a primary consideration; each pound of warhead weight required approximately 50 pounds in gross lift-off weight (mostly in the form of propellants, but also in the structure needed to support the propellants).[56]

Unfortunately, graphite is subject to rapid oxidation and vaporization at high surface temperatures, which often led to structural failures. Eventually, protective coatings to isolate the material from the airstream would be developed as part of the Dyna-Soar and Space Shuttle programs, but this was much too late to help the original ballistic missile programs.[57]

---

53. This was at odds with what Charles McClellan and John Becker found at Langley when Inconel X was selected as the primary material for the hot structure of the X-15.
54. Stalder, "The Useful Heat Capacity of Several Materials for Ballistic Nose-Cone Construction," p. 6.
55. Ibid., pp. 7, 19.
56. Ibid., pp. 1–2; Chuck Hansen, *U.S. Nuclear Weapons: The Secret History* (Arlington, TX: Aerofax, Inc., 1988), pp. 195–196.
57. Stalder, "The Useful Heat Capacity of Several Materials for Ballistic Nose-Cone Construction," pp. 7–8.

The blunt-body theory led directly to the Mk I[58] reentry vehicle developed by General Electric for the intermediate-range Thor and early Atlas-B/C[59] ballistic missiles. Despite the weight penalty, General Electric selected 1,200 pounds of copper for the Mk I as a matter of expediency, since the material was well understood and readily available. A slightly improved Mk II reentry vehicle initially used a similar amount of copper.[60]

General Electric had decided against beryllium because its low ductility posed severe fabrication difficulties, and the company lacked the time to develop new processes to manufacture the metal. The Navy, on the other hand, had more time and selected a beryllium heat sink for the initial Mk 1 warheads[61] for its Polaris submarine-launched ballistic missile. Based on the success of the Navy program, beryllium soon replaced copper in later versions of the Air Force Mk II warhead.

The Air Force Mk II reentry vehicle, and its similar, but unrelated, Navy Mk 1 counterpart, had a low ballistic coefficient and a large heat sink. Although both of these vehicles offered an immediate solution to the reentry problem, experience quickly showed that heat-sink warheads could fail in several ways. For instance, they could lose their strength because of the high temperatures and disintegrate under the air and deceleration loads. Alternately, they could fail mechanically due to thermal stresses or fail materially through spalling,[62] melting, or sublimation. At the extreme, the warhead could combust due to an unstable exothermic reaction of the material with the airstream.

Even when the early warheads worked as designed, they had significant defects as weapon delivery systems. For instance, they spent a lot of time in the upper atmosphere, trailing a stream of ionized gas from the melting heat sink that showed up on radar, which made them very susceptible to interception by antiballistic missiles. In addition, decelerating in the upper atmosphere

---

58. As frequently happens within the USAF, the Mk I, II, and III reentry vehicle designations used Roman numerals, while later devices used Arabic numbers.

59. Atlas was designed as an intercontinental missile, but the early versions (A/B/C/D) were closer in performance to an intermediate-range missile. The later E/F versions were true intercontinental missiles.

60. Stalder, "The Useful Heat Capacity of Several Materials for Ballistic Nose-Cone Construction," pp. 1–2; Hansen, *U.S. Nuclear Weapons*, pp. 195–196.

61. The USAF and Navy used similar nomenclature for their reentry vehicles, but an Air Force Mk I was decidedly different from a Navy Mk 1 (and note, the Navy used Arabic numbers from the beginning).

62. Spall are flakes of a material that break off a larger solid body and can be produced by a variety of mechanisms, including projectile impact, corrosion, and weathering. Spalling and spallation both describe the process of surface failure in which spall is shed.

meant that the warhead was traveling relatively slowly when it reached the lower atmosphere, making it susceptible to winds and frequently causing it to miss its target by miles.

Engineers soon discovered that accuracy could be improved by increasing the descent velocity so the reentry vehicle was less affected by winds. They had come full-circle as what they now wanted was a vehicle with a high ballistic coefficient, or exactly the opposite of the blunt-body concept that had made reentry possible in the first place. This, however, increased the heat load past the absorption capability of a heat sink of any practical size, so researchers began investigating methods to supplement or replace the metallic heat sinks.[63]

Thermal Protection Systems

In the broadest sense, there are three thermal protection concepts: passive, semipassive, and active. The type of protection on any space-venturing vehicle or, more precisely, on any given area of a vehicle depends largely on the magnitude and duration of the heat load as well as various operational considerations.

As the name implies, passive thermal protection systems have no moving parts. They are the simplest but, until the advent of the Space Shuttle, had the least capability. These concepts have fallen into three general categories: heat sink, hot structure, and insulated structure. The heat sink absorbs almost all of the incident heat and stores it in a large, usually metallic mass. Additional mass may be added to increase the heat storage capability, but in general the concept is limited to short heat pulses. A hot structure allows the temperature to rise until the heat being radiated from the surface is equal to the incident heating, much like the heating element of an electric stove. This concept is not limited by the duration of the heat pulse but is restricted to the acceptable surface temperature of the structural material. The Inconel X hot structure of the X-15 research airplane could withstand temperatures up to about 1,200 °F, which was about the maximum temperature for the concept. Insulated structures use an outer shell that radiates most incident heat away from an underlying structure protected by a layer of some insulating material, usually high-temperature ceramic-fiber batt insulation. Both the magnitude and duration of the heat pulse are limited for insulated systems, but it allows lower-temperature structural materials to be used.

---

63. Leonard Roberts, "A Theoretical Study of Nose Ablation," and Aleck C. Bond, Bernard Rashis, and L. Ross Levin, "Experimental Nose Ablation," both in "NACA Conference on High-Speed Aerodynamics," pp. 253–284.

There are two basic semipassive concepts. Heat pipes are attractive where there is a localized area of high heating with an adjacent area of low heating. A working fluid vaporizes in the high-heat area, and the vapor flows naturally to a cooler region where it condenses and the heat is rejected. The condensed working fluid is returned to the high-heat region by capillary action.

The other semipassive concept is ablation. Ablation is a process in which a material (ablator) sacrifices itself to protect the underlying structure. However, the ablator is consumed in the process, thus limiting the duration of its operation and generally eliminating reuse. These ablative materials may be chemically constructed (usually some form of fiberglass or a spray-on resin-based coating) or made from natural materials (both the Chinese and Soviets/Russians used oak wood on some early reentry vehicles). As heat is absorbed by the ablator, part of the material decomposes into a gas, which carries heat away from the surface and into the boundary layer where the chemical byproducts further block convective heating. Near the surface, exposure to higher temperatures causes more complete pyrolysis, thus the surface shows a more complete decomposition than the inner layers and forms a char, much like charcoal.[64]

The vaporization of an ablator is an endothermic reaction; i.e., it needs energy input to proceed. A proper ablator will absorb the heat flux incident to its surface and vaporize at a rate proportional to the magnitude of the heat flux. The heat is carried away with the vaporized material, increasing the effectiveness of the thermal protection system. The vaporization leads to a thinning of the ablator, resulting in an upper limit on the amount of heat it can absorb. Perhaps the most unappreciated advantage of ablation is that the coolant (the gas generated by pyrolysis), after accepting all of the heat it is capable of absorbing, is automatically jettisoned. It requires no complex pumps or other mechanical devices, saving weight and complexity. In addition, as the ablated material is jettisoned, the ensuing heat load is lessened by the continuous reduction of mass.[65] This was particularly important for the early ballistic missiles and human space programs for which weight was critical due to launch vehicles with limited capability.[66]

Three different active cooling concepts have been widely investigated. Transpiration and film cooling operate on a principle similar to the ablators;

64. R. Bryan Erb and Stephen Jacobs, "Entry Performance of the Mercury Spacecraft Heat Shield," a paper presented to the Heat Protection Session of the AIAA Entry Technology Conference, October 12–14, 1964, p. 5.

65. H. Julian Allen, "The Aerodynamic Heating of Atmosphere Entry Vehicles."

66. A.V. Levy, "Evaluation of Reinforced Plastics Material in High Speed Guided Missiles and Power Plant Application," *Plastics World* 14 (March 1956): 10–11.

coolant ejected from the vehicle surface blocks most of the heat from reaching the underlying structure. These concepts use a pump to bring liquid coolant from a remote reservoir onto the surface of the vehicle. Transpiration cooling ejects the coolant through a porous surface, whereas film cooling uses discrete slots to flow coolant along the vehicle parallel to the airflow. The mass penalties associated with expendable coolant, of course, have usually limited these concepts to small, high-heat regions. The last concept—one first envisioned by Wernher von Braun and pioneered by Bell Aircraft—circulated coolant (water or liquid metal) around the hot area then through a heat exchanger or radiator, much like the engine-cooling system on an automobile. This water-wall concept allowed almost any heat load to be tolerated, but the coolant and radiator systems added considerable weight and were therefore not truly practicable.

Better Warheads

The General Electric Mk I and Mk II heat sink passive reentry warheads provided the United States with an immediate method to field intermediate-range Thor and Atlas-B/C ballistic missiles. However, their limitations prevented their use on warheads intended for the longer-range ICBMs such as Titan and the improved Atlas-D/E/F. Avco Manufacturing Corporation (now part of Textron) and General Electric, along with numerous military and NACA laboratories, began investigating other types of thermal protection systems for these warheads. Initially, many researchers thought the answer could be found in the active cooling concepts like those assumed by Allen and Eggers for the blunt-body theory, but these proved to be heavy and unreliable, particularly when subjected to the decelerations experienced by the ICBM reentry vehicles.

While the U.S. Air Force and its contractors were expending efforts on heat sinks, the U.S. Army was taking the lead in investigating ablators. At Redstone Arsenal, AL, the Vitro Corporation was using the exhaust from liquid rocket engines as a heat source to test a new family of ablators. On August 8, 1957, a Jupiter-C launched from Cape Canaveral carrying a subscale reentry vehicle to an altitude of 600 miles and a range of 1,200 miles and proved the feasibility of the ablative-type nose cone during reentry.[67] It should be noted, however, that this test was only meant to validate the reentry conditions for an intermediate-range Jupiter warhead. Whether an ablative ICBM warhead would work was still unproven, although in theory the concept could be scaled

---

67. Swenson, Greenwood, and Alexander, *This New Ocean*, p. 64.

up for those more advanced systems. The Air Force initially looked upon the Army's claimed success with skepticism.[68]

The NACA was involved from the beginning in the warhead research, and it contributed to both the heat sink and the ablative experiments. Several NACA researchers that worked on the early problems of reentry would go on to have great influence on the human space flight program. Maxime A. Faget, Paul E. Purser, Robert O. Piland, and Robert R. Gilruth would all play important roles in future programs. Faget and Purser were also members of the Polaris Task Group, which provided advice to the U.S. Navy on the Polaris warheads, and the task group worked extensively with Lockheed in developing the Navy Mk 1 heat-sink warheads for the Polaris A1 and A2.[69]

NACA researchers also investigated ablative materials such as Teflon, nylon, and fiberglass using new test facilities. At Langley, these included an acid-ammonia rocket facility that could generate a maximum temperature of 4,100 °F and a gas velocity of 7,000 ft/sec, and an ethylene-air jet facility capable of yielding temperatures of 3,500 °F. Electric arc-jet facilities were built at both Langley and the Lewis Research Center, and each facility was capable of generating temperatures up to 12,000 °F. The NACA researchers proved the Army was right: ablative coatings were superior to heat sinks.[70]

General Electric saw the promise of ablators, resulting in the ablative Mk III warhead for the intermediate-range Jupiter-C, Thor, and Atlas-D missiles. The ablative heat shield consisted of nylon cloth impregnated with a phenolic plastic resin. High-temperature ablators were an important discovery for the mid-ballistic-coefficient warheads then under development and paved the way for the high-ballistic-coefficient warheads used today. Avco entered the fray with the ablative Mk 4 for the Atlas-E/F and the even longer range Titan I,[71] as well as the ablative Mk 5 for the early Minuteman I. All future warheads used ablative heat shields as their ballistic coefficients became higher and their velocity during reentry increased.[72]

On the improved Mk 6 reentry vehicle for the Titan II, the use of a nylon phenolic ablator was so effective that significantly reduced bluntness was possible, with a half-angle of only 12.5 degrees. The 10-foot-long Mk 6 was the largest warhead developed by the United States, with an entry mass of

---

68. Hansen, *U.S. Nuclear Weapons*, p. 195.

69. Swenson, Greenwood, and Alexander, *This New Ocean*, p. 65.

70. Ibid., pp. 64–65.

71. "Mark 4 Operational Re-Entry Vehicle Titan/Atlas Quarterly Program Progress Report," Avco Corporation, Operational Missiles Subdivision, Report 21-138.1, May 10, 1961.

72. Hansen, *U.S. Nuclear Weapons*, p. 195.

A Titan II intercontinental ballistic missile with an Mk 6 reentry vehicle. USAF.

7,500 pounds. Subsequent advances in nuclear weapon design allowed reentry vehicles to become significantly smaller, and the Avco Mk 11 warhead began the shift toward a more pointed body with an even higher ballistic coefficient.

Eventually, the reentry vehicles used by the later long-range ballistic missiles (such as the Minuteman III, Trident, and Peacekeeper) returned to the long, slender, pointed vehicles that had long been associated with space travel. New ablative coatings protected these reentry vehicles during the plunge through the atmosphere, minimizing the time the vehicle spent in the upper atmosphere and lowering its chances of detection and interception as well as significantly reducing the accuracy dispersions due to upper-altitude winds. The General Electric Mk 12 reentry system for the Minuteman III, deployed in 1970, contained three separate warheads, each having its own pointed reentry vehicle. The Avco Mk 21 replaced the Mk 12 on the Air Force Peacekeeper (MX), and the Navy adopted a version called the Mk 5 for the Trident D-5. Typical of the latest warheads, the sharp-nose Mk 21 had a fine-weave, pierced-carbon-fabric nose tip followed by a graphite-epoxy body that was covered with an ablative carbon-phenolic heat shield.[73]

The CORONA Satellite Reconnaissance Program

Even as these activities were taking place, in early 1956 President Dwight D. Eisenhower authorized a new satellite reconnaissance project jointly managed by the Central Intelligence Agency and the U.S. Air Force. Its intention was to place in polar orbit, with all possible speed, a satellite carrying a camera that could take photographs of the Soviet Union (as well as other nations), return the film to Earth, and provide spatial and visual data to analysts on which to build more reliable intelligence estimates for decision makers. The long shadow of the successful surprise attack on Pearl Harbor in 1941 had prompted U.S. intelligence organizations to expand their reconnaissance efforts to avoid being caught unaware again.

Because of the successful effort during the 1950s to develop a reconnaissance satellite—Project CORONA had its first successful flight on August 18, 1960, after several failures—a new era of intelligence gathering began. To disguise its true purpose, it was given the cover name "Discoverer" and described as a scientific research program. This highly classified reconnaissance

---

73. Ibid., pp. 195–202. Some sources, including Hansen, indicate the Mk 12 multiple independently targetable reentry vehicle (MIRV) was also used on the Minuteman II, but this is unlikely given the throw-weight of the earlier missile. This is confirmed in Daniel Ruchonnet, "MIRV: A Brief History of Minuteman and Multiple Reentry Vehicles," Lawrence Livermore Laboratory Report COVD-1571 (February 1976).

effort—that acquired 3,000 feet of film with coverage of over 1,650,000 square miles of the former Soviet Union—revolutionized how the United States collected and used foreign intelligence. It ushered in an era in which the intelligence community regularly received imagery that offered a synoptic view of much of Earth's surface. The intelligence community thereafter had both a high volume and continuous flow of data from satellite imagery. CORONA also was an important milestone from a historical perspective. It was the first imaging reconnaissance satellite, the first source of mapping imagery from space, the first source of stereo imagery from space, the first space program to succeed with multiple reentry vehicles, and the first space reconnaissance program to fly 100 missions.[74]

The CORONA program used a unique film-return capsule, essentially a version of the ICBM reentry vehicle, to reenter the atmosphere and return safely to Earth. An ablative heat shield, like those developed earlier for the ICBM program, dissipated the energy on the spacecraft as it entered the atmosphere and slowed down; then the heat shield was jettisoned and a parachute was deployed that enabled an Air Force plane to capture the capsule as it followed its trajectory into the Pacific Ocean. From August 1960 to May 1972, more than 120 successful CORONA missions provided invaluable intelligence on the Soviet Union and other nations.

CORONA was succeeded by a series of evermore sophisticated reconnaissance satellites, and a continuous stream of data began to flow from its imagery. However, it was the last system that used film imagery that had to be physically returned to Earth for processing and analysis. Overseen by the National Reconnaissance Office, the more than 50-year-old satellite reconnaissance effort has been enormously significant. Politicians, intelligence professionals, and the general public all view it as critical to the welfare of the United States and its continued national sovereignty. It enables the discovery of strategic weapons, military buildups, and troop movements, and it provides independent verification of strategic weapons-reduction efforts. President Lyndon Baines Johnson did not overestimate the importance of the CORONA satellite reconnaissance technology in 1967 when he said, in light

---

74. The best works on this subject include Dwayne A. Day, John M. Logsdon, and Brian Latell, eds., *Eye in the Sky: The Story of the Corona Spy Satellite* (Washington, DC: Smithsonian Institution Press, 1998); Robert McDonald, ed., *CORONA: Between the Sun & the Earth: The First NRO Reconnaissance Eye in Space* (Bethesda, MD: American Society for Photogrammetry and Remote Sensing, 1997); Curtis Peebles, *The Corona Project: America's First Spy Satellites* (Annapolis, MD: Naval Institute Press, 1997); Philip Taubman, *Secret Empire: Eisenhower, the CIA, and the Hidden Story of America's Space Espionage* (New York: Simon & Schuster, 2003).

During the early years of satellite reconnaissance, spacecraft ejected film canisters that reentered the atmosphere and were retrieved by aircraft. This CORONA return canister is being recovered by a C-119 snatching its parachute during descent in 1962. USAF.

of the fact that the United States probably spent between $35 and $40 billion (in 1960s dollars) on it, "If nothing else had come of it except the knowledge we've gained from space photography, it would be worth 10 times what the whole program has cost."[75] Accordingly, this mission will continue indefinitely, and the United States will invest in numerous future reconnaissance satellite programs throughout the 21st century.

At the same time, Allen and Egger's concept of reentry would soon become important for a strikingly different use: NASA's human space flight program initiated with the creation of the Agency on October 1, 1958. The application of reentry and recovery technology to the NASA human space flight program is the subject of the next chapter.

---

75. Quoted in *The NRO at the Crossroads: Report of the National Commission for the Review of the National Reconnaissance Office* (Washington, DC: National Reconnaissance Office, November 1, 2000), Appendix E, p. 120.

Walt Disney and Wernher von Braun teamed up in the 1950s to create the expectation that space flight was a near-term possibility with spaceplanes. NASA.

CHAPTER 2

Human Space Flight and the Problem of Reentry

The limitations of the heat-sink approach to returning to Earth, pioneered for warhead reentry, became even more apparent when the method was considered for a piloted spacecraft. First, a returning orbital spacecraft would enter at a faster velocity than a ballistic missile and would get much hotter, requiring more copper or beryllium. Perhaps even more importantly, flying a piloted capsule into orbit and back home, and recovering a fragile human pilot, is more difficult than was lobbing a warhead at Moscow.

However, there was an even greater problem. The heat sink itself got hot, and some of this heat—possibly several hundred degrees—transferred to the vehicle. This was a workable issue for a nuclear weapon but prohibitive both for the film-return canister used on the first reconnaissance program, CORONA, and on a piloted spacecraft.

Although almost all of the early development into reentry was oriented toward warheads, a small number of researchers believed the efforts were also applicable to future piloted spacecraft. In fact, the X-15 research airplane used a manifestation of the blunt-body concept during its suborbital reentry. True ballistic reentry, although acceptable for missile warheads, produced g-loads far too high (80 g's or more) for human survival. However, by using the shallower reentry trajectory made possible by the blunt-body theory, the g-loads and peak temperatures could be significantly reduced, and the early ablative heat shields being developed for warheads provided a solution that made a small human space capsule possible. The g-loads (approximately 8 g's for about 90 seconds) were still a concern, but centrifuge studies showed these were within the limits of human tolerance if the crew was in a reclining position and the force was applied from front to back (eyeballs-in). However, studies of the X-15 and Dyna-Soar cockpits produced primarily eyeballs-down g-loads. Both problems required serious efforts to understand reentry for human pilots. This led to piloted capsules designed so crews would lie on their backs, facing away from the direction of flight.

These concerns for reentry and recovery from space for a piloted spacecraft were far removed from the public discussion of space flight in the 1950s before

Sputnik's launch on October 4, 1957. For example, in 1951 when Willy Ley, a former member of the German VfR and himself a skilled promoter of space flight, organized a Space Travel Symposium at the Hayden Planetarium in New York City, the speakers hardly mentioned the problems of returning from space. Out of this symposium came the celebrated series of *Collier's* articles appearing between 1952 and 1954 that did much to popularize space flight in President Dwight D. Eisenhower's America. An editorial suggested that space flight was possible and was not just science fiction, and that it was inevitable that humanity would venture outward. It framed the exploration of space in the context of the Cold War rivalry with the Soviet Union and concluded the following:

> *Collier's* believes that the time has come for Washington to give priority of attention to the matter of space superiority. The rearmament gap between the East and West has been steadily closing. In addition, nothing, in our opinion, should be left undone that might guarantee the peace of the world. It's as simple as that.[1]

There was nothing in any of this to suggest that returning from space was in any manner whatsoever difficult or would require extended experimentation. Wernher von Braun led off the *Collier's* issue with an impressionistic article describing the overall features of an aggressive space flight program. He advocated the orbiting of an artificial satellite to learn more about space flight followed by the first orbital flights by humans; he also advocated developing a reusable spacecraft for travel to and from Earth's orbit, building a permanently inhabited space station, and, finally, human exploration of the Moon and planets by spacecraft launched from the space station. Ley and several other writers followed with elaborations on various aspects of space flight ranging from technological viability to space law to biomedicine.[2] The series concluded with a special issue of the magazine devoted to Mars in which von Braun and others described how to get there and predicted what might be found based on recent scientific data.[3] It was all going to be so simple and straightforward. Practical study and experimentation would soon prove otherwise.

---

1. "What Are We Waiting For?" *Collier's* (March 22, 1952), p. 23.

2. "Man Will Conquer Space Soon" series, *Collier's* (March 22, 1952), pp. 23–76ff.

3. Wernher von Braun with Cornelius Ryan, "Can We Get to Mars?" *Collier's* (April 30, 1954), pp. 22–28.

The X-Planes

Despite the obvious advances in rocket technology, before humans could venture into space, it would first be necessary to fly much faster than the 500 mph possible at the end of World War II. Since the beginning of powered flight, wind tunnels had proven to be useful tools, but they had several limitations regarding high-speed research. Signifying perhaps the greatest limitation, in the 1930s it became apparent that the transonic regime could not be adequately simulated due to the physical characteristics of the wind tunnel test sections. The only alternative was to use real airplanes.

This led directly to what became known as the Round One X-planes, so called because the Air Force assigned the aircraft an X (for experimental) designation.[4] On October 14, 1947, Air Force Captain Charles E. Yeager became the first human to break the sound barrier in level flight when the Bell X-1[5] achieved Mach 1.06 at 43,000 feet. It took 6 additional years before NACA test pilot A. Scott Crossfield exceeded Mach 2 in the Navy/Douglas D-558-2 Skyrocket on November 20, 1953. Only 3 years later, on September 7, 1956, Air Force Captain Milburn G. Apt was killed during his first X-2 flight after he reached Mach 3.196 (1,701 mph), becoming the first person to fly at three times the speed of sound, albeit briefly.[6]

The original rationale behind the X-planes was to explore a flight regime that the wind tunnels could not simulate. However, by the time the X-1 and D-558 actually flew, researchers had figured out how to extend ground test facilities into this realm with the slotted-throat wind tunnel, developed by Richard Whitcomb at the NACA Langley Memorial Aeronautical Laboratory in the early 1950s. Therefore, the real value of the research airplanes lay in the comparison of the ground-based techniques with actual flight results to validate theories and wind tunnel results. The fact that the first transonic flights showed nothing particularly unexpected—dispelling the myth of the sound barrier—was of great relief to the researchers. The fearsome transonic zone had

---

4. Although several X-planes explored the high-speed/altitude flight environment, many others investigated less glamorous areas and received substantially less publicity.
5. Technically designated XS-1 (experimental, supersonic) at the time, this would be changed to X-1 in 1948.
6. Jay Miller, *The X-Planes: X-1 to X-45* (Hinckley, U.K.: Midland Counties Publishing, 2001), pp. 9–11; Richard P. Hallion, *On The Frontier: Flight Research at Dryden* (Washington, DC: NASA SP-4303, 1984).

John V. Becker and the 11-Inch Hypersonic Tunnel at Langley Aeronautical Laboratory in 1950. NASA EL-2002-00243.

been reduced to an ordinary engineering problem.[7] Although few people were thinking about it at the time, the results from these experiments would also be instrumental in developing the spaceplanes that had long been discussed.

The next step was to push through the so-called thermal barrier predicted by Theodore von Kármán and others. Although not related to specific velocity, like the speed of sound, vehicles venturing above Mach 5 (hypersonic velocities) experienced significantly increased heating rates that appeared to present a significant problem. Between the two World Wars, hypersonics had been an area of theoretical interest to a small group of researchers, but little progress was made in defining the possible problems and even less toward solving them. The major constraint was propulsion. Engines, even the rudimentary rockets then being experimented with, were incapable of propelling any significant object to hypersonic velocities. Wind tunnels also lacked the power to generate such speeds. Computer power to simulate the environment using models, what is now known as computational fluid dynamics, had not yet even been imagined.

Hypersonic research was authorized primarily to support the massive effort associated with developing intercontinental missiles.[8] One researcher interested

---

7. John V. Becker, *The High-Speed Frontier: Case Studies of Four NACA Programs, 1920–1950* (Washington, DC: NASA SP-445, 1980), pp. 93–94.

8. Memorandum for the Record, Minutes of the NACA Subcommittee on Stability and Control, June 4, 1951, NASA Historical Reference Collection.

in exploring the new science of hypersonics was John V. Becker, then the assistant chief of the Compressibility Research Division at NACA Langley. On August 3, 1945, Becker proposed the construction of a new type of supersonic wind tunnel for Mach 7. A number of uncertainties caused Becker to suggest building a small pilot tunnel with an 11-by-11-inch test section to demonstrate the new concept. In September 1945, Charles H. McLellan and a small staff began work on the facility, and the first test, on November 26, 1947, revealed uniform flow at Mach 6.9, essentially meeting all the original intents.[9] Nevertheless, many researchers wanted to build an actual hypersonic vehicle to validate the data from the new test facilities. The large rocket engines being developed for the missile programs were seen as possible powerplants for a hypersonic research vehicle. It was time for the Round Two X-planes.[10]

Round Two X-Planes

In late August 1954, researchers at NACA Langley released a 4½-page paper entitled "NACA Views Concerning a New Research Airplane," giving a brief background on the hypersonic research airplane and attaching a study by Becker as a possible solution. The researchers at Langley also proposed a secondary mission to explore space-related concerns, particularly reentry techniques, but most other researchers dismissed this because they believed human space flight was many decades off. Nevertheless, the paper listed two major areas to be investigated: (1) preventing the destruction of the aircraft structure by the direct or indirect effect of aerodynamic heating, and (2) achieving stability and control at very high altitudes, at very high speeds, and during atmospheric reentry from ballistic flightpaths. They were important goals on the way toward human space flight.[11]

An industry competition in 1955 resulted in the Air Force awarding North American Aviation a contract to build three experimental X-15 hypersonic research airplanes. The Government-industry team—led by Becker, Hartley A. Soulé, and Walter C. Williams from the NACA, and Crossfield, Charles H.

---

9. John V. Becker to the Langley Chief of Research, "Proposal for New Type of Supersonic Wind Tunnel for Mach Number 7.0," August 3, 1945, NASA Historical Reference Collection; Becker, "Results of Recent Hypersonic and Unsteady Flow Research at the Langley Aeronautical Laboratory," pp. 619–628; James R. Hansen, *Engineer in Charge: A History of the Langley Aeronautical Laboratory, 1917–1958* (Washington, DC: NASA SP-4305, 1987), p. 347.

10. Becker, "The X-15 Program in Retrospect," p. 2.

11. "NACA Views Concerning a New Research Airplane," August 1954; Becker, "The X-15 Program in Retrospect," pp. 2–3.

Feltz, and Harrison A. "Stormy" Storms, Jr., for North American—would soon become the stuff of legend.[12] North American had accepted an extraordinarily difficult task when the company agreed to develop the new hypersonic research airplane. Eventually, some 2,000,000 engineering work-hours and over 4,000 wind tunnel hours were devoted to finalizing the X-15 configuration.[13]

Researchers at Langley considered two basic structural design approaches. The first was a conventional low-temperature design of aluminum or stainless steel protected from the high-temperature hypersonic environment by a layer of assumed, but as-yet undeveloped, insulation. The other was an exposed hot structure in which the materials and design approach would permit high structural temperatures and would not require protection from the induced environment. It was similar, in many regards, to the heat-sink solution for missile warheads.[14]

Heating projections made by researchers for various trajectories showed that the airplane would need to accommodate an equilibrium temperature of 2,000 °F on its lower surface. At the time, no known insulating material could meet this requirement. The most likely candidate was the Bell double-wall technique in which a low-temperature structure was protected by a high-temperature outer shell with some insulation in between. This concept would later undergo extensive development in support of Dyna-Soar, but in 1954 it was in an embryonic stage and, in any case, was not applicable to the critical nose and leading-edge regions that would experience much of the heat load.[15]

It was by no means obvious that the hot-structure approach would prove practical either. The permissible design temperature for the best available materials was far below the peak 2,000 °F equilibrium temperature during reentry. It was clear that either direct internal cooling or heat absorption into the structure itself would be necessary, but each approach was expected to bring a heavy weight penalty.

---

12. Crossfield had left NACA to join North American in order to be more closely associated with the X-15 program. When NASA and the USAF decided that all record flights would be flown by Government research pilots, this effectively left Crossfield out of the record books, but he stated on numerous occasions that he felt he made the correct decision.

13. Harrison Storms, "The X-15 Rollout Symposium," October 15, 1958, released statements in the files at the AFFTC History Office.

14. Becker, "The X-15 Project, Part I: Origins and Research Background," *Astronautics and Aeronautics II* (February 1964): 56–57. These same trade studies would be repeated many times during the concept definition for Space Shuttle.

15. Ibid. Possible insulators included water, several different liquid metals, air, and various fibrous batt materials. The liquids would require active pumps and large reservoirs, making them exceptionally heavy concepts.

Shock waves festoon a small-scale model of the X-15 in NASA Langley's 4-by-4-foot Supersonic Pressure Tunnel. NASA L-1962-02577.

Exotic alloys made from the rare earth elements were still laboratory curiosities, so the list of candidate materials narrowed to corrosion-resistant steels (CRES) and magnesium-, titanium-, and nickel-base alloys. Although AM-350 CRES and 6A1-4V titanium had good strength over a wide temperature range, the strength of both alloys tended to fall off rapidly above 800 °F. Fortunately, various magnesium and nickel alloys exhibited a gradual drop in strength up to 1,200 °F, and Inconel X[16]—for all intents an exotic stainless steel—was ultimately chosen for the outer skin of the airplane. In a happy coincidence,

---

16. Inconel X® is a temperature-resistant alloy whose name is a registered trademark of Huntington Alloy Products Division, International Nickel Company, Huntington, WV. Inconel X is 72.5 percent nickel, 15 percent chromium, and 1 percent columbium, with iron making up most of the balance.

researchers at Langley discovered that the skin thickness needed to withstand the expected aerodynamic stresses was about the same needed to absorb the thermal load. This meant that it was possible to solve the structural problem for the transient conditions of the hypersonic research aircraft with no serious weight penalty for heat absorption.[17]

Other items developed for the X-15 included one of the first stable platforms (what is today called an inertial measurement unit), auxiliary power units, a ball nose to measure local flow direction in the high-temperature airstream (usually, incorrectly, called a Q-ball), reaction control rockets for attitude control in the space environment, and workable full-pressure suits. Various incarnations of all of these would find use on future spacecraft.[18] Perhaps even more important was the development of extensive engineering and mission-simulation systems. Although crude by current standards, the X-15 pioneered the use of simulators not just to train pilots but also to engineer the aircraft, plan the missions, and understand the results. Not surprisingly, given the involvement of Feltz, Storms, and Williams in the subsequent Apollo and Space Shuttle programs, the X-15 pointed the way to how America would conduct its human space missions. Simulation is one of the enduring legacies of the small, black experimental airplanes.

The X-15 flight program encompassed 199 flights between June 1959 and October 1968. By July 1962, the X-15 had flown more than Mach 6 and above 300,000 feet, exceeding its design goals. NASA test pilot Joseph A. Walker would fly the maximum-altitude flight of the program, 354,200 feet, on August 22, 1963. This was his second excursion above 62 miles in just over a month, making him the first person to venture into space (albeit suborbital space) twice. Air Force Major William J. "Pete" Knight flew the maximum-speed mission, 4,520 mph (Mach 6.70) on October 3, 1967.[19]

---

17. Richard L. Schleicher, "Structural Design of the X-15," North American Aviation (1963), pp. 13–14.

18. For more detail on all of these systems, see Dennis R. Jenkins and Tony R. Landis, *Hypersonic: The Story of the North American X-15* (North Branch, MN: Specialty Press, 2003).

19. James E. Love, "History and Development of the X-15 Research Aircraft," n.d., pp. 15–16; Joseph A. Walker, "Pilot Report for Flight 3-21-32," July 19, 1963; Johnny G. Armstrong to Dennis R. Jenkins, August 3, 2002; and Robert G. Hoey to Dennis R. Jenkins, August 12, 2002, all in the personal possession of Dennis R. Jenkins. Bob White would also be the first person to fly Mach 5 and Mach 6, the first to fly above 200,000 and 300,000 feet, and he set a Federation Aeronautique Internationale altitude record of 314,750 feet in the X-15 that still stands in 2011. Although Walker was the first person to make two space flights, it took until August 2005 for NASA to recognize him as an astronaut.

Neil Armstrong is seen next to X-15-1 (56-6670) after a research flight. Armstrong made his first X-15 flight on November 30, 1960. This was the first X-15 flight to use the ball nose, which provided accurate measurement of flow angle (attitude) at supersonic and hypersonic speeds. NASA E60-6286.

The X-15s amassed a large body of hypersonic flight data that researchers are still using more than 50 years later. One of the largest contributions from the program was in conclusively demonstrating that there were no hypersonic facilities barriers. Becker remembered that many aerodynamicists had expected that because of strongly interacting flow fields, viscous interactions with strong shock waves, and possible real-gas effects, the X-15 would reveal large discrepancies between wind tunnel data and actual flight. Instead, the X-15 demonstrated that most flight results agreed substantially with low-temperature, perfect-gas wind tunnel predictions. Data from the X-15, coupled with data from the ASSET and PRIME programs, allowed the design of future hypersonic vehicles, including the Space Shuttle, to proceed with much greater confidence.[20]

---

20. Becker, "The X-15 Program in Retrospect"; Kenneth W. Iliff and Mary F. Shafer, "A Comparison of Hypersonic Vehicle Flight and Prediction Results," NASA TM-104313 (October 1995), pp. 5–6.

When Project Mercury began, it rapidly dominated some of the research areas that had first interested X-15 planners, such as microgravity studies and reaction-control systems pioneered in this program and incorporated whole-sale into the Mercury capsule. These reaction controls effectively maintained attitude control in space for both the X-15 and the Mercury capsule, and, beyond that, the X-15 furnished valuable information on the blending of reaction controls with conventional aerodynamic surfaces during exit and reentry—a matter of concern to subsequent Space Shuttle development. The X-15 clearly demonstrated the ability of pilots to fly rocket-powered aircraft out of the atmosphere and back and to make precision landings. The director of the Flight Research Center, Paul F. Bikle, saw the X-15 and Mercury as a "parallel, two-pronged approach to solving some of the problems of piloted space flight. While Mercury was demonstrating humanity's capability to func-tion effectively in space, the X-15 was demonstrating man's ability to control a high-performance vehicle in a near-space environment…considerable new knowledge was obtained on the techniques and problems associated with lifting reentry." They were important steps toward an eventual spaceplane.[21]

Faget and Nonlifting Reentry

Throughout most of 1957, research into human space flight continued to examine winged vehicles. In particular, the NACA cooperated with the Air Force on the Manned Glide Rocket Research System, and the Ames laboratory looked into a flattop, round-bottom skip-glider. Nevertheless, after Sputnik was launched on October 4, 1957, it became obvious that a ballistic capsule was both the best and quickest way to get Americans into orbit; the available launch vehicles simply could not support the increased weight of a winged vehicle. Even so, what would later become Dyna-Soar was being discussed, culminating with the Round Three X-plane conference at Ames on October 15, 1958.[22]

A minority contingent within the NACA, mainly at Langley, continued to argue that lifting reentry vehicles would be far superior to a nonlifting capsule. In fact, at the NACA Conference on High-Speed Aerodynamics in March 1958, Becker presented a concept for a piloted 3,060-pound winged orbital satellite. According to Becker, this concept drew more industry reaction—almost all of

---

21. Wendell H. Stillwell, *X-15 Research Results* (Washington, DC: NASA SP-60, 1965), p. iv; Walter C. Williams, "The Role of the Pilot in the Mercury and X-15 Flights," in the *Proceedings of the Fourteenth AGARD General Assembly,* September 16–17, 1965, Portugal.

22. Swenson, Greenwood, and Alexander, *This New Ocean*, p. 71.

it favorable—than any other paper he had written, including the initial X-15 study. What ruled out acceptance of his proposal was that the 1,000 pounds of extra weight (compared to the capsule design presented by Maxime A. Faget) was beyond the capability of the Atlas launch vehicle. If the Titan had been further along, the concept would have worked, and Becker believed that the first U.S. piloted spacecraft might well have been a landable winged vehicle.[23]

Although wings brought many benefits during entry and landing, they also brought structural weight, and there was no practical, and little theoretical, data applicable to the development of a spaceplane. On the other hand, Faget at NACA Langley believed the first piloted spacecraft should be a simple nonlifting capsule that could take full advantage of the research and production experience gained through development of warheads for the ballistic-missile programs. Faget thought that much of the aerodynamic and thermodynamic work done by the NACA on ballistic-missile warheads was directly applicable to the piloted ballistic capsule. Given a maximum weight of about 2,000 pounds, a vehicle with a base diameter of 7 feet would provide sufficient volume for a single occupant and the required equipment.[24]

Faget opined that the capsule concept had certain attractive operational aspects including following a ballistic path that minimized requirements for the autopilot, guidance, and control systems. In his mind, this not only resulted in a significant weight savings but also eliminated the hazard of malfunction. The ballistic capsule only had to perform a single maneuver to orient itself properly and then fire its retrorockets. Faget pointed out that this did not need to be done with a great deal of precision for the vehicle to successfully enter the atmosphere, but he did acknowledge the disadvantage of a large-area landing by parachute with no corrective control during entry.[25]

There was, of course, another disadvantage. At an entry angle of –0.8 degree, the vehicle would endure approximately 8.5 g's on the way down (although the buildup was expected to be relatively gradual). Faget believed this was well within the tolerances of the test pilots expected to inhabit the vehicle, and he noted that larger entry angles with even more severe deceleration should be tolerable with the proper training and equipment. He readily accepted this as a program requirement.[26]

---

23. Hansen, *Engineer in Charge*, pp. 377–381, Becker quote from p. 381. The subject was confirmed in a telephone conversation between the author and Becker, June 17, 2001.
24. Maxime A. Faget, Benjamin J. Garland, and James J. Buglia, "Preliminary Studies of Manned Satellites—Wingless Configuration: Nonlifting," NACA RM L58E07a (August 11, 1958), p. 4.
25. Ibid., pp. 1–2.
26. Ibid., p. 3.

Researchers investigated four different shapes for the vehicle: a hemisphere, a heavily blunted 15-degree cone, a 53-degree cone, and a nearly flat nose with a spherical segment. Perhaps the greatest unknown was the effect of laminar flow on the vehicle. Limited testing on rocket-boosted, free-flying models at Wallops Island confirmed that laminar flow could be expected on at least the flat-nose shape. If laminar flow was truly present, the total heat load could be reduced by an order of magnitude over the ballistic-missile warheads.[27] Faget concluded his study by saying: "It appears that, insofar as reentry and recovery is concerned, the state of the art is sufficiently advanced so that it is possible to proceed confidently with a piloted satellite project based upon the ballistic reentry type of vehicle."[28]

Even before the first Mercury flight, however, it was recognized that a purely ballistic vehicle was not the ideal approach. In addition to g-loads and heating during reentry, the poor predictability of the final impact point forced the use of a large contingent of ships spread over a wide area of ocean for recovery. Ultimately, Mercury was the only U.S. piloted spacecraft to use a purely ballistic reentry without some ability to steer the vehicle.[29]

Providing a small amount of lift—an aerodynamic force perpendicular to the flightpath—reduced the severity of reentry and improved the accuracy of recovery. For the capsules, lift could be created by slightly offsetting the center of gravity from the vertical axis. The trajectory could then be changed by rolling the vehicle about its vertical axis so that the offset lift vector could be pointed in any direction. The roll maneuver could be accomplished via a reaction wheel (gyroscope) or by using small reaction-control thrusters, which is the approach that ultimately was selected. The lift could also be directed upward to maintain a small flightpath angle for as long as possible. Once the vehicle passed the high-speed heating pulse, it could be banked to turn toward the desired recovery point, producing a relatively small cross-range capability.[30] A lift-to-drag ratio (L/D) of only 0.2 could cut the maximum deceleration in half and stretch the range by 280 miles.[31]

---

27. Ibid., p. 5.

28. Ibid., pp. 6–7.

29. Maxime A. Faget et al., "Preliminary Studies of Manned Satellites—Wingless Configurations: Nonlifting," in "NACA Conference on High-Speed Aerodynamics," p. 25.

30. D.J. Lickly et al., "Apollo Reentry Guidance," Massachusetts Institute of Technology (MIT) Instrumentation Laboratory Report R-415 (July 1963); Robert G. Hoey, *Testing Lifting Bodies at Edwards* (Palmdale, CA: PAT Projects, Inc., September, 1994).

31. Faget, Garland, and Buglia, "Preliminary Studies of Manned Satellites—Wingless Configuration: Nonlifting," NACA RM L58E07a, pp. 6–7.

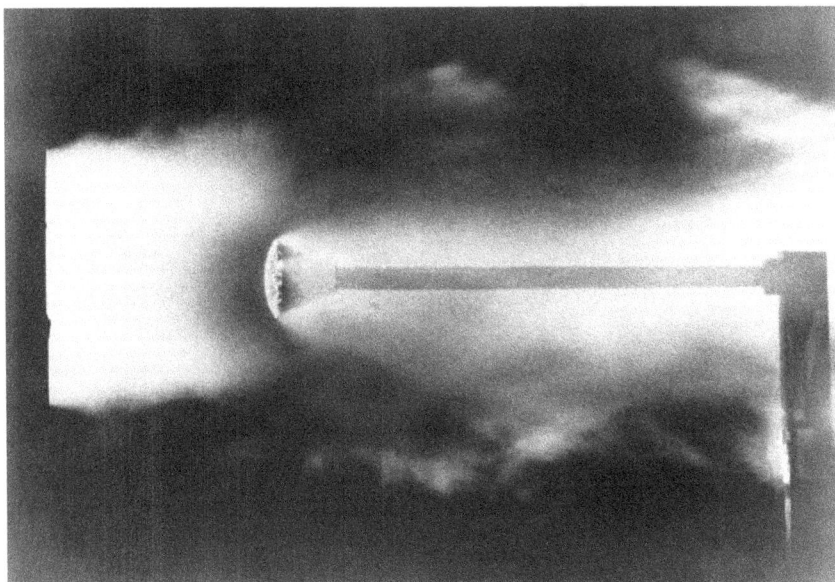

A hot jet research facility, used extensively in the design and development of the reentry heat shield on the Project Mercury spacecraft. The electrically heated arc jet simulates the heating encountered by a space vehicle as it returns to Earth's atmosphere at high velocities. NASA EL-2002-00308.

The resulting semiballistic reentries were of longer duration than a pure ballistic reentry, but they produced more tolerable g-loads and lower peak temperatures. A more complex design was needed for the ablative heat shield because the heating was not symmetrical and the duration of the heat pulse was longer, driving the development of a nonreceding, charring ablator that stayed in place to maintain the outer mold line and aerodynamic character-istics of the vehicle. Gemini and Apollo used semiballistic capsules protected by nonreceding, charring ablators and were initially intended to be recovered on land. Ultimately, however, both vehicles still required large ocean areas and significant recovery forces at the end of their missions.[32]

Although not a design factor during the early space race, all of these ballistic and semiballistic vehicles were designed to survive only a single reentry. The only capsule known to have flown twice was Gemini II, which after its initial test flight was modified as a prototype Gemini-B for the Manned Orbiting Laboratory (MOL). An entirely new heat shield was used for the second flight on a developmental Titan IIIC, on November 3, 1966.

---

32. *Testing Lifting Bodies at Edwards.*

Man in Space Soonest

While the NACA was pursuing its studies for a human space flight program, the U.S. Air Force proposed the development of a piloted orbital spacecraft under the title of "Man In Space Soonest" (MISS).[33] Initially discussed before the launch of Sputnik 1, in October 1957, the Air Force invited Dr. Edward Teller and several other leading members of the scientific/technological elite to study the issues of human space flight and make recommendations for the future. Teller's group concluded that the Air Force could place a human in orbit within 2 years and urged the department to pursue reasons to undertake this effort. Teller understood, however, that there was essentially no military reason for the mission and chose not to tie his recommendation to any specific rationale, falling back on a basic belief that the first nation to put a man in space would gain national prestige and generally advance in science and technology.[34] Soon after the new year, Lieutenant General Donald L. Putt, the Air Force Deputy Chief of Staff for Development, informed NACA Director Hugh L. Dryden of the Air Force's intention to pursue aggressively "a research vehicle program having as its objective the earliest possible manned orbital flight which will contribute substantially and essentially to follow-on scientific and military space systems." Putt asked Dryden to collaborate in this effort, with the NACA taking on the role of a decidedly junior partner.[35] Dryden agreed, but by the end of the summer the newly created NASA was leading the U.S. human space flight effort, and the Air Force was the junior player.[36]

Notwithstanding the lack of a clear-cut military purpose, the Air Force pressed for MISS throughout the first part of 1958, clearly expecting to become the lead agency in any U.S. space program. Specifically, it believed hypersonic spaceplanes and lunar bases would serve well the country's national security

---

33. The Man-in-Space-Soonest program called for a four-phase capsule orbital process, which would first use instruments, to be followed by primates, then a pilot, with the final objective of landing humans on the Moon. See David N. Spires, *Beyond Horizons: A Half Century of Air Force Space Leadership* (Peterson AFB, CO: Air Force Space Command, 1997), p. 75; Swenson, Grimwood, and Alexander, *This New Ocean*, pp. 33–97.

34. Swenson, Grimwood, and Alexander, *This New Ocean*, pp. 73–74.

35. Lt. Gen. Donald L. Putt, USAF Deputy Chief of Staff, Development, to Hugh L. Dryden, NACA Director, January 31, 1958, Folder #18674, NASA Historical Reference Collection.

36. NACA to USAF Deputy Chief of Staff, Development, "Transmittal of Copies of Proposed Memorandum of Understanding between Air Force and NACA for joint NACA–Air Force Project for a Recoverable Manned Satellite Test Vehicle," April 11, 1958, Folder #18674, NASA Historical Reference Collection.

needs in the coming decades. To help make that a reality, the Air Force requested $133 million for the MISS program and secured approval for the effort from the Joint Chiefs of Staff.[37] Throughout this period, a series of disagreements between Air Force and NACA officials aggravated both sides. The difficulties reverberated all the way to the White House, prompting a review of the roles of the two organizations.[38] Dryden, the normally staid and proper director of the NACA, complained in July 1958 to the President's science advisor, James R. Killian, about the lack of clarity regarding the role of the Air Force versus the NACA. He asserted, "The current objective for a manned satellite program is the determination of man's basic capability in a space environment as a prelude to the human exploration of space and to possible military applications of manned satellites. Although it is clear that both the National Aeronautics and Space Administration and the Department of Defense should cooperate in the conduct of the program, I feel that the responsibility for and the direction of the program should rest with NASA." He urged that the President state a clear division between the two organizations on the human space flight mission.[39]

As historians David N. Spires and Rick W. Sturdevant pointed out, the MISS program became derailed within the DOD at essentially the same time because of funding concerns and the lack of a clear military mission:

> Throughout the spring and summer of 1958 the Air Force's Air Research and Development Command had mounted an aggressive campaign to have ARPA convince administration officials to

---

37. The breakdown for this budget was aircraft and missiles—$32M; support—$11.5M; construction—$2.5M; and R&D—$87M. See Memorandum for Advanced Research Projects Agency (ARPA) Director, "Air Force Man-in-Space Program," March 19, 1958, Folder #18674, NASA Historical Reference Collection.

38. Maurice H. Stans, Director, Bureau of the Budget, Memorandum for the President, "Responsibility for 'Space' Programs," May 10, 1958; Maxime A. Faget, NACA, Memorandum for Dr. Dryden, June 5, 1958; Clotaire Wood, Headquarters, NACA, Memorandum for files, "Tableing [sic] of Proposed Memorandum of Understanding Between Air Force and NACA For a Joint Project For a Recoverable Manned Satellite Test Vehicle," May 20, 1958, with attached Memorandum, "Principles for the Conduct by the NACA and the Air Force of a Joint Project for a Recoverable Manned Satellite Vehicle," April 29, 1958; Donald A. Quarles, Secretary of Defense, to Maurice H. Stans, Director, Bureau of the Budget, April 1, 1958, Folder #18674, all in NASA Historical Reference Collection.

39. Hugh L. Dryden, Director, NACA, Memorandum for James R. Killian, Jr., Special Assistant to the President for Science and Technology, "Manned Satellite Program," July 19, 1958; Folder #18674, NASA Historical Reference Collection.

approve its Man-in-Space-Soonest development plan. But ARPA [Advanced Research Projects Agency] balked at the high cost, technical challenges, and uncertainties surrounding the future direction of the civilian space agency.[40]

Project Mercury

President Eisenhower signed the National Aeronautics and Space Act of 1958 into law at the end of July and the next month assigned the nation's human space flight mission to NASA. Thereafter, the MISS program was folded into what became Project Mercury.[41] By early November 1958, DOD had acceded to the President's desire that the human space flight program be a civilian effort under the management of NASA. For its part, NASA invited Air Force officials to appoint liaison personnel to the Mercury program office at Langley, and they did.[42]

Unlike the spaceplane concept pursued by the Air Force during the early planning for MISS, the Mercury spacecraft configuration was primarily determined by heating considerations, and the final shape evolved from an extensive study of how to minimize the effects of reentry heating while providing adequate stability.[43] The first configuration, designed mostly by Faget, used a conical body with a flat heat shield—essentially a beefed-up, piloted Mk II

---

40. David N. Spires and Rick W. Sturdevant, "'...to the very limit of our ability...': Reflections on Forty Years of Civil-Military Partnership in Space Launch," in Roger D. Launius and Dennis R. Jenkins, eds., *To Reach the High Frontier: A History of U.S. Launch Vehicles* (Lexington: University Press of Kentucky, 2002), p. 475.

41. For an overall discussion of the early military human program, see Dwayne A. Day, "Invitation to Struggle: The History of Civilian-Military Relations in Space," in John M. Logsdon, with Dwayne A. Day and Roger D. Launius, eds., *Exploring the Unknown: Selected Documents in the History of the U.S. Civil Space Program, Volume II, External Relationships* (Washington, DC: NASA SP-4407, 1996), pp. 248–251.

42. Memorandum for Dr. Silverstein, "Assignment of Responsibility for ABMA Participation in NASA Manned Satellite Project," November 12 1958; Abe Silverstein to Lt. Gen. Roscoe C. Wilson, USAF Deputy Chief of Staff, Development, 20 November 1958; and Hugh, L. Dryden, Deputy Administrator, NASA, Memorandum for Dr. Eugene Emme for NASA Historical Files, "The 'signed' Agreement of April 11, 1958, on a Recoverable Manned Satellite Test Vehicle," September 8, 1965, Folder #18674, all in NASA Historical Reference Collection.

43. Erb and Jacobs, "Entry Performance of the Mercury Spacecraft Heat Shield," p. 2.

NASA undertook tests of full-size Mercury capsules (either launched from the ground by rocket power or dropped from airplanes at high altitude) to test the capsule's dynamic stability and aerodynamic heating as well as the effectiveness of the pilot-escape and parachute-recovery systems. This test took place in 1958. NASA EL-2000-00282.

warhead. The afterbody was recessed slightly from the perimeter of the heat shield to minimize heat transfer to the inhabited part of the spacecraft. Tests at Langley showed this configuration was unstable at subsonic speeds and suffered various heating problems during reentry, leading to the afterbody becoming more squat and extending all the way to the perimeter of the heat shield. The flat heat shield was soon discarded because it trapped too much heat, leading to a rounded bottom, and NASA researchers decided on a 1.5 ratio between the radius of the curve and the diameter of the heat shield. The final obstacle was figuring out where to locate the landing parachutes. Eventually, engineers decided to add a cylinder on top of the afterbody (the frustum), creating the now familiar Mercury shape.[44]

Mercury is the only U.S. piloted spacecraft that used a purely ballistic entry, with conditions dictated primarily by the g-limits for the pilot. The

---

44. Swenson, Greenwood, and Alexander, *This New Ocean*, p. 95.

anticipated environment included a heat pulse of approximately 8,910 BTU per square foot for a spacecraft weighing 2,439 pounds. Temperatures in the shock layer immediately ahead of the blunt-body heat shield were expected to be in excess of 10,000 °F. To provide adequate protection for the astronaut, whose back was only a few inches away from the inner surface of the heat shield, engineers set the maximum permissible temperature on the back of the heat shield at 150 °F. In addition to handling the expected head load, the heat shield needed to withstand the dynamic pressures expected during reentry as well as relatively high water-impact forces while the heat shield was still hot. The heat shield also needed to withstand the acoustic and vibration environments imposed during launch and ascent as well as the hard vacuum and low temperatures of the space environment.[45]

The Air Force, along with Avco Corporation and General Electric, recommended using an ablative heat shield, based on the later Mk II reentry vehicle. NASA researchers, however, worried that the relatively recent ablative technology was too immature, and they wanted to stay with a heat sink similar to the initial General Electric Mk II design.[46] A slightly refined reentry profile reduced the total heat load to approximately 6,000 BTU per square foot, and this could be accommodated by a 1-inch-thick slab of beryllium with a unit weight of 10 pounds per square foot. This forged heat sink was fabricated by sintering beryllium powder, a process generally similar to the one used to manufacture the early Mk II reentry vehicles.

Ultimately, NASA researchers concluded that the toxicity and the fire hazard posed by the beryllium heat sink in the event of a nonwater landing was unacceptable. However, by the time this decision was made, a number of heat sinks had been manufactured and, mostly to provide an early flight capability, were used on the suborbital flights of Alan Shepard and Virgil I. "Gus" Grissom. Once NASA had decided that beryllium posed unacceptable risks, they again turned to the ballistic missile researchers for an answer.[47]

Ablation technology was being rapidly advanced in the quest for a medium ballistic-coefficient reentry vehicle for the long-range Atlas and Titan ICBMs. In fact, an ablative-covered Jupiter reentry vehicle was successfully recovered in the fall of 1958, just as the Mercury spacecraft was entering preliminary design. The Jupiter reentry vehicle used a composite design that provided structural strength while allowing large volumes of hot gasses to flow between the ablative

---

45. Erb and Jacobs, "Entry Performance of the Mercury Spacecraft Heat Shield," p. 6.

46. Swenson, Greenwood, and Alexander, *This New Ocean*, pp. 70, 95.

47. Erb and Jacobs, "Entry Performance of the Mercury Spacecraft Heat Shield," p. 2.

Detail B

Typical cross-section
of afterbody skin

Outside skin

Airspace

Insulation

Inside skin

0.3"

1.5"

Afterbody skin
(see detail B)

Heat shield (see detail A)

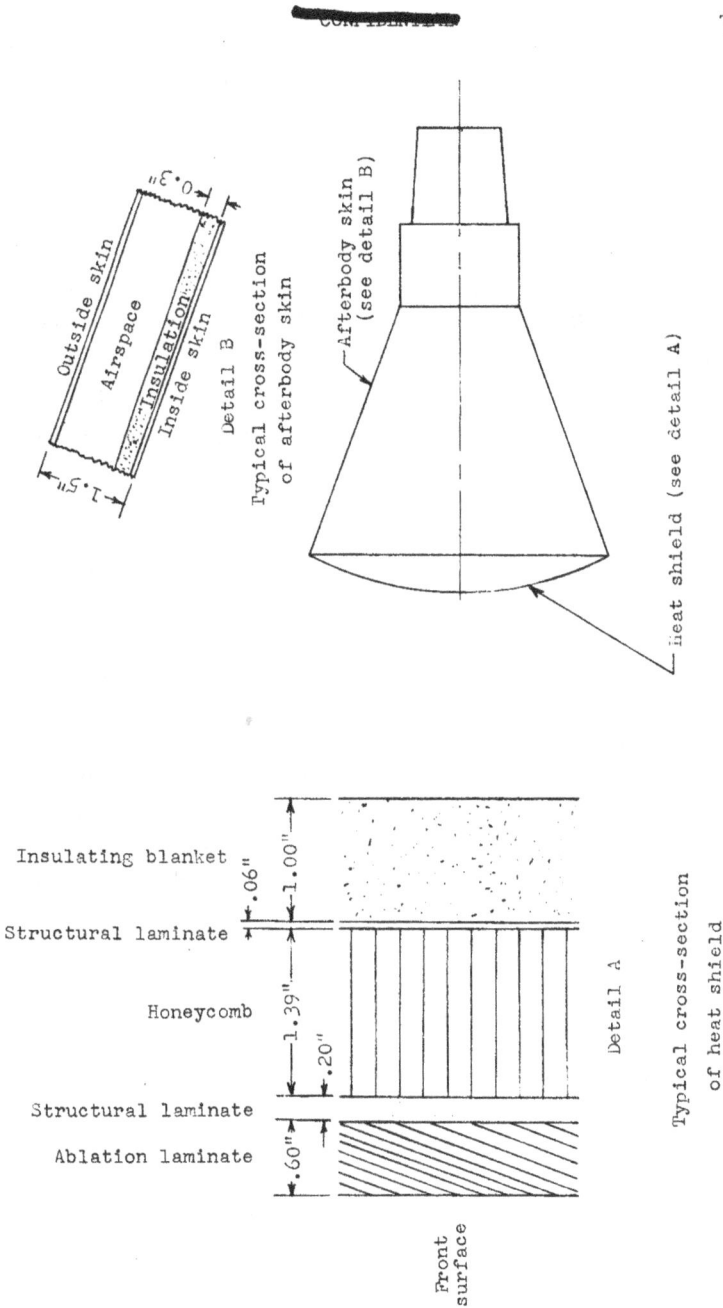

Figure 7-15.- Structural details of ablation shield.

Insulating blanket

Structural laminate

Honeycomb

Structural laminate

Ablation laminate

.06"

1.00"

1.39"

.20"

.60"

Detail A

Typical cross-section
of heat shield

Front
surface

Structural details of the Mercury heat shield. NASA.

laminates without forcing the laminates to separate. The design appeared ideal for Mercury, and NASA quickly adopted it to replace the beryllium heat sink.[48]

Unfortunately, the Jupiter reentry vehicle used a trajectory more closely resembling the suborbital Mercury flights instead of the anticipated orbital missions. No directly applicable reentry experience existed, so NASA researchers wanted to conduct a flight qualification test to demonstrate that the ablative heat shield could withstand an orbital entry. The mechanics of orbital flight were still being developed, so researchers decided a ballistic (suborbital) mission would pose less risk and would better ensure that the heat shield could be recovered for postflight analysis.[49] A carefully chosen flight profile would largely duplicate the heating rates and total heat load of an orbital entry. NASA tested this with the "Big Joe" flight program consisting of two flights, each using an Atlas launch vehicle that would loft a boilerplate Mercury capsule 1,500 miles downrange.[50]

Although the outer mold line was identical, the structure of the boilerplates for the Big Joe tests was not typical of production Mercury capsules. For instance, the boilerplate only contained a half-size pressure vessel to support instrumentation rather than the full-size pressurized cabin contoured to the outer mold line. The 2,555-pound boilerplate was built in two segments, with the lower blunt body manufactured at the Lewis Research Center and the upper afterbody manufactured at Langley, both under the direction of the Space Task Group. The main monocoque of the afterbody was fabricated using thin sheets of corrugated Inconel. General Electric supplied the outer heat shield, with parts of the interior structure manufactured by B.F. Goodrich. The heat shield was instrumented with 51 thermocouples to obtain temperature and char-penetration time histories during flight. The boilerplate did not carry a retro-package or an escape tower, but it did contain a parachute recovery system.[51]

The Big Joe heat shields consisted of two parts: an outer ablation laminate that was 1.075 inches thick and an inner structural laminate that was 0.550 inch thick. The ablation laminate was made of concentric layers of fiberglass cloth oriented so that the layers were at a 20-degree angle to the local heat shield surface. The structural laminate was made of similar fiberglass cloth oriented with the individual layers parallel to the outer surface. Both the ablation and structural

---

48. Ibid.

49. Ibid.

50. Ibid.; Swenson, Grimwood, and Alexander, *This New Ocean*, p. 201.

51. Robert L. O'Neal and Leonard Rabb, "Heat-Shield Performance During Atmospheric Entry of Project mercury Research and Development Vehicle," NASA-TM-A-490 (May 1961), pp. 3–4; Swenson, Grimwood, and Alexander, *This New Ocean*, p. 201.

Engineers inspect a boilerplate Mercury space capsule with an array of bulky test equipment strewn about. NASA GPN-2000-003008.

layers were made from a special finish fiberglass cloth and a phenolic resin. The resin content of the ablation and structural laminates was, by weight, 40 percent and 30 percent, respectively. A 3-inch-high circular ring made from fiberglass and resin was attached to the back of the heat shield and was used to bolt the heat shield to the pressurized compartment of the boilerplate.[52]

The first test flight, Big Joe 1, was launched on Atlas 10-D (628) from the Air Force Missile Test Center Launch Complex 14 at Cape Canaveral on September 9, 1959. The Atlas was programmed to rise, pitch over horizontally as it reached its 100-mile peak altitude, and then pitch down slightly before releasing the boilerplate at a shallow angle downward. The Atlas had three engines: a sustainer engine in the middle and two booster engines strapped to the sides of the main missile. At a predetermined time during ascent, the two booster engines were designed to separate from the missile, as their thrust no longer compensated for their additional weight and drag. A malfunction on Atlas 10-D kept this from happening, and the additional weight caused the

---

52. Erb and Jacobs, "Entry Performance of the Mercury Spacecraft Heat Shield," p. 3.

sustainer engine to deplete its propellants 14 seconds early and delay the separation of the boilerplate from the launch vehicle. Nevertheless, the 13-minute flight reached an altitude of 90 miles and traveled 1,424 miles downrange. The boilerplate reached a maximum velocity of 14,857 mph, a maximum dynamic pressure (max-q) of 675 pounds per square foot, and sustained a maximum of 12 g's. The parachutes operated successfully, and the USS *Strong* (DD-758) recovered the boilerplate in excellent condition off Puerto Rico.[53]

Because of the launch vehicle anomaly, the peak-heating rate obtained during the flight was only 77 percent of that expected, and the total cold-wall heat load was only 42 percent of the expected 7,100 BTU per square foot.[54] Despite not meeting the initial objectives, the flight was generally successful. The recovered boilerplate showed the heat shield withstood both the reentry and recovery operations with only superficial damage. The heat shield exhibited generally uniform heating, and the visibly charred region penetrated to a depth of approximately 0.20 inch, or some 12 percent of the total thickness. Small cracks and areas of delamination were found, but these did not extend in depth beyond the visibly charred portion and did not affect the structural integrity of the heat shield. Analysis showed no measurable profile change, and the heat shield lost only 6.1 pounds of mass due to the ablation process.[55]

The results from the flight test confirmed that the theories (models) being used by the researchers and engineers were valid and would satisfactorily predict heating for other trajectories. This gave NASA researchers enough confidence to cancel the Big Joe 2 test (Atlas 20-D), which had been scheduled for the fall of 1959, and the launch vehicle was transferred to another program. The Big Joe 1 boilerplate is currently displayed at the Smithsonian National Air and Space Museum's Steven F. Udvar-Hazy Center at Chantilly, VA.

In addition to the Big Joe and Wallops Island tests, materials for the Mercury heat shield were evaluated in several NASA test facilities. Models ranging from full scale to 1 percent of full scale were used. Researchers conducted the tests in plasma-arc, radiant-lamp, and oxyhydrogen blowtorch facilities. In general, these tests yielded poor results because of contamination in the atmosphere (usually helium) and poor control over the heating rates. Fortunately, the Langley Structures Division researchers devised a way to test the heat shield

---

53. 45th Space Wing Launch Database, courtesy of Mark Cleary at Patrick AFB; Erb and Jacobs, "Entry Performance of the Mercury Spacecraft Heat Shield," p. 3.

54. Erb and Jacobs, "Entry Performance of the Mercury Spacecraft Heat Shield," p. 3.

55. O'Neal and Rabb, "Heat-Shield Performance During Atmospheric Entry of Project Mercury Research and Development Vehicle"; Erb and Jacobs, "Entry Performance of the Mercury Spacecraft Heat Shield," p. 4.

An Atlas launch vehicle carrying the Big Joe capsule leaves its launch pad on a 2,000-mile ballistic flight to an altitude of 100 miles on September 1, 1959. This boilerplate model of the orbital capsule was recovered and studied for the effect of reentry heat and other flight stresses. NASA MSFC-9139360.

in their new arc-jet facility using air as a working medium, eliminating the contamination issues associated with the earlier tests. This facility could not accommodate a full-size heat shield, so samples were cut from a production heat shield (no. 13) that had been rejected due to defects discovered during ultrasonic inspection. These tests further characterized the behavior of the ablative heat shield and confirmed the adequacy of the theories and models used to develop it.[56]

Most of the afterbody of the Mercury capsule was protected by a series of René 41 shingles with fibrous batt insulation underneath. The beaded (corrugated) shingles were 0.016 inch thick and were attached by bolts through oversized holes that allowed the shingles to expand and contract without buckling. Oversized washers covered the holes to minimize heat and air penetration. René 41 consists of 53 percent nickel, 19 percent chromium, 11 percent cobalt, 9.75 percent molybdenum, 3.15 percent titanium, 1.6 percent aluminum, 0.09 percent carbon, 0.005 percent boron, and less than 2.75 percent iron. Other parts of the afterbody were covered with beryllium shingles fabricated from hot-pressed beryllium blocks. Thermoflex RF insulation blankets were located between these shingles and the primary structure. Both the René 41 and beryllium shingles were coated on the outer surface with blue-black ceramic paint to enhance their radiative properties. The inner surface of the beryllium shingles had a very thin gold coating to attenuate thermal radiation into the spacecraft.[57]

In October 1959, Mercury prime contractor McDonnell received the first production Mercury ablative heat shield from General Electric. This unit was used on the first Mercury capsule, which was delivered to Wallops Island on April 1, 1960, for a beach-abort test that took place on May 9.[58] Surprisingly, the production heat shields were slightly thinner than the Big Joe unit, with an ablation laminate only 0.65 inch thick and a structural laminate 0.30 inch thick. Several problems were encountered during the fabrication of the production heat shields, most concerning whether the desired inclination of the laminates was being maintained. Ultimately, bore samples showed that it was not possible to keep the inclination correct near the center of the heat shield, raising a concern that this area would delaminate during reentry. Engineers from Langley, General Electric, and McDonnell concluded that the best course of action was to machine a 15-inch diameter section out of the center of each

56. Erb and Jacobs, "Entry Performance of the Mercury Spacecraft Heat Shield," pp. 7–8.

57. P.W. Malik and G.A. Souris, "Project Gemini," NASA CR-1106 (June 1968), pp. 13–14.

58. "Mercury," available online at *http://www.astronautix.com/project/mercury.htm*, accessed August 23, 2009.

Recovery practice for the Mercury program after landing in the ocean conducted by the USS *Strong* (DD-758) in 1959. NASA.

heat shield and replace it with a molded plug that was layed-up separately (and correctly, since the fabric did not need to stretch the entire diameter of the heat shield). The plug was secured in the heat shield by 12 inclined dowel pins and a layer of glue.[59]

The background provided by the single Big Joe flight and the arc-jet tests gave NASA researchers confidence that the operational heat shield would perform satisfactorily. Since the suborbital Redstone launches had used capsules equipped with beryllium heat sinks, the first real use of an ablative heat shield was on the Mercury-Atlas 2 (MA-2) test flight on February 21, 1961. The flight was designed to subject the spacecraft (no. 6) to maximum g-loads (15.9 g's) and to produce the maximum afterbody heating; consequently, the total heat load on the heat shield was less than the load during a nominal orbital entry. MA-2 flew a successful suborbital mission that lasted 17 minutes, 56 seconds and reached an altitude of 114 miles and a speed of 13,227 mph. The capsule

---

59. Erb and Jacobs, "Entry Performance of the Mercury Spacecraft Heat Shield," p. 7.

was recovered 1,432 miles downrange, and subsequent analysis showed the heat shield performed satisfactorily.[60]

The operational experience with the Mercury heat shield was generally satisfactory, but the center plug section continued to cause problems. During the first orbital flight of the Mercury heat shield on the MA-4, on September 13, 1961, the center plug section cracked free at the outer diameter and the resulting gap was about 0.1875 inch, although the inclined dowel pins still tightly retained the plug.[61] The flight of Enos the Chimp in MA-5 on November 20, 1961, ended with the center plug completely missing from the heat shield. In this case, the holes used by the dowels had been drilled too shallow, allowing the pins to slip free, and air bubbles were discovered in the glue. A transducer under the center plug showed that the plug had been in place up to the instant of water impact, and there was no damage to Enos or the capsule.[62]

McDonnell and General Electric subsequently developed improved inspection techniques, and all of the remaining heat shields were inspected and repaired as necessary to ensure the center plug did not separate on future missions. Despite the precautions, the center plug was again missing after the first U.S. piloted orbital mission of John Glenn in MA-6 on February 20, 1962, and after the MA-7 flight of Scott Carpenter on May 24, 1962.[63]

When Wally Schirra was launched in MA-8 on October 3, 1962, the materials and construction of the heat shield were the same as the previous orbital missions with the exception that the center plug was bolted to the structural laminate to prevent its loss after entry. During the postflight examination, engineers found the center plug was still firmly attached. The heat shield provided excellent protection during entry, and the stagnation point appeared to have been very close to the center of the heat shield, as expected of a ballistic (nonlifting) entry. The most worrisome finding was evidence of separation where the ablation laminate was glued to the structural laminate. Based on the depth of the char around the cracks, the separation appeared to have developed after the peak heating, perhaps as late as water impact (the ablator was still quite hot when it hit the cold ocean). The char depth measured 0.33 to 0.44 inch, about the same as previous missions. The measured weight loss, however, was 17.43 pounds—substantially more than the 13 pounds on previous missions. Engineers noted that the technique used for drying the

---

60. Ibid., p. 8.

61. Ibid.

62. Ibid.

63. Ibid.

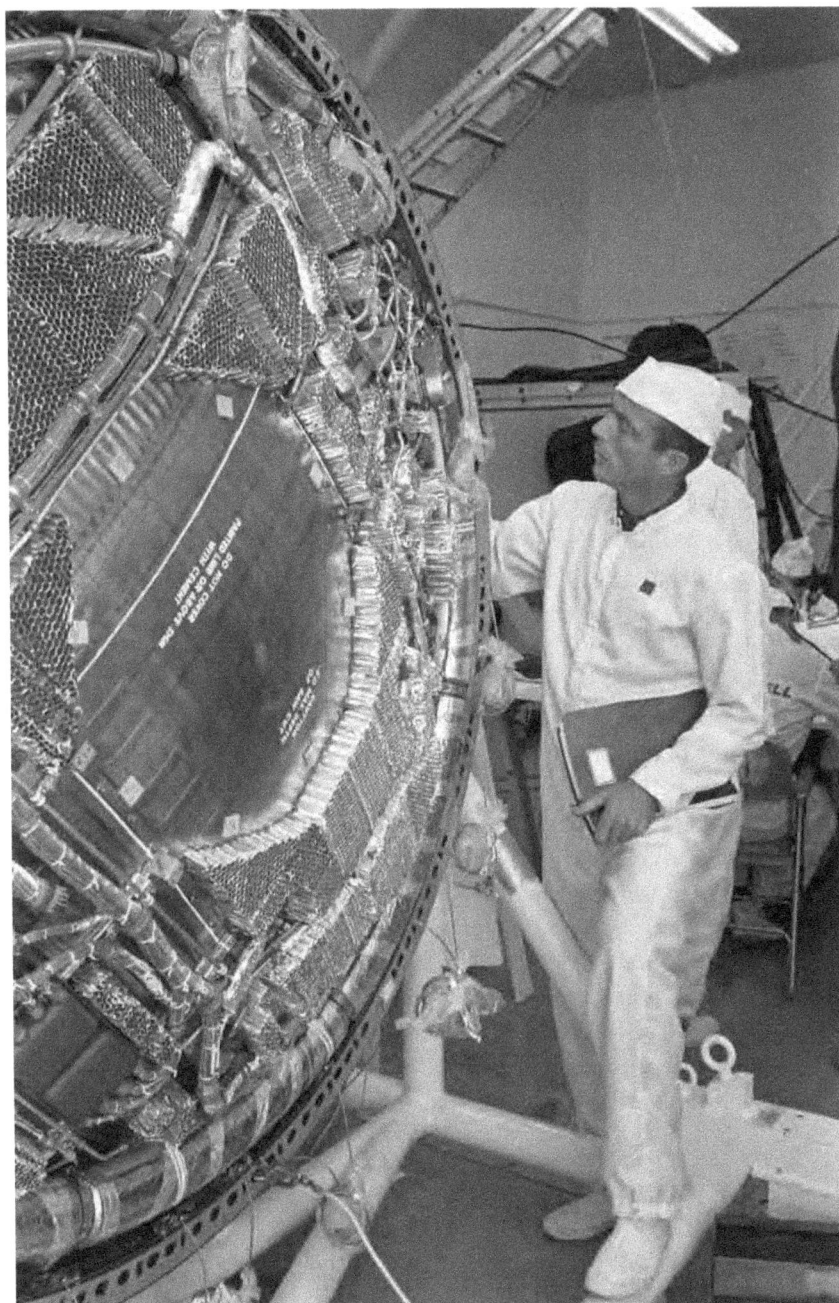

Inside Hangar S at Cape Canaveral, Mercury astronaut M. Scott Carpenter examines the honeycomb protective material on the main pressure bulkhead (heat shield) of his Mercury capsule, nicknamed "Aurora 7." NASA S62-01420.

heat shield prior to weighing it changed for MA-8, so the results might not be directly comparable.[64]

The last Mercury flight was piloted by Gordon Cooper in MA-9 on May 14 to 15, 1963. Concerns about the delamination on MA-8 led NASA to install 14 bolts through the ablative laminate into the structural laminate to keep them from separating if the glue failed completely. This was a fortunate decision as postflight inspection revealed that the glue had separated, but the heat shield stayed together thanks to the bolts. Using the same drying procedure as on MA-8, engineers noted the heat shield lost 15.34 pounds.[65]

Project Gemini

Announced on December 7, 1961, by Robert R. Gilruth, Gemini was the third civilian U.S. human space flight program to be approved, after both Mercury and Apollo.[66] Designed to fill the gap between Mercury and Apollo, Gemini was originally called Mark II, referring to its early status as a larger and slightly improved Mercury capsule. In the end, it would prove to be much more, although it bore a family resemblance to Mercury. The most significant of the proposed changes, which was never implemented, was the Rogallo wing paraglider recovery system, which is discussed in chapter 3.

One of the objectives for Gemini was to demonstrate a controlled reentry to a preselected landing site. Although the Gemini spacecraft had the same general shape as the ballistic Mercury capsule it was derived from, including the blunt heat shield and rotational symmetry about its longitudinal axis, it had an inherent lifting capability produced by a vertical center-of-gravity displacement with respect to the longitudinal axis. This center-of-gravity displacement introduced a trim angle of attack that, because of the blunt heat shield, produced a lift vector in the opposite direction from the trim angle. The vehicle could be rolled about the relative velocity vector or stability axis to provide a small maneuvering capability during reentry. If the vector modulation was done precisely enough, the spacecraft could be guided to a preselected

---

64. Ibid., p. 9.

65. Ibid., pp. 9–10.

66. Barton C. Hacker and James M. Grimwood, *On the Shoulders of Titans: A History of Project Gemini* (Washington, DC: NASA SP-4206, 1977), pp. 1–3. Mercury was the first approved program (September 1958), followed by Apollo (September 1960). Separately, the Air Force Dyna-Soar (December 1957) program predated all of the NASA projects.

RESEARCH CONTRIBUTING TO PROJECT MERCURY

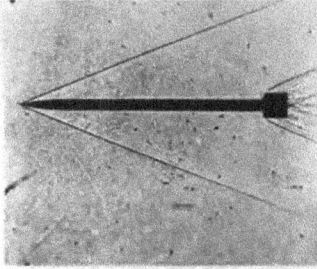

INITIAL CONCEPT

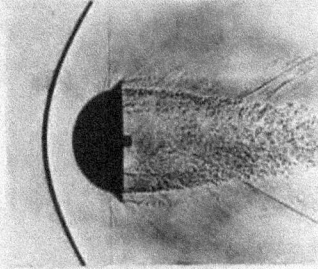

BLUNT BODY CONCEPT 1953

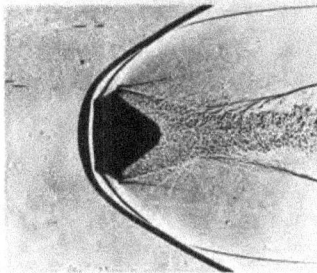

MISSILE NOSE CONES 1953-1957

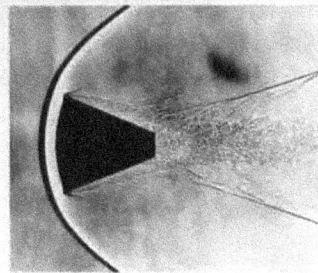

MANNED CAPSULE CONCEPT 1957

These four shadowgraph images represent early reentry vehicle concepts. A shadowgraph is a process that makes visible the disturbances that occur in a fluid flow at high velocity. Light passing through a flowing fluid is refracted by the density gradients in the fluid, resulting in bright and dark areas on a screen placed behind the fluid. A blunt body, as pioneered by H. Julian Allen, produces a shock wave in front of the vehicle—visible in the photo—that actually shields the vehicle from excessive heating. As a result, blunt-body vehicles can stay cooler than pointy, low-drag vehicles can. NASA 60-108.

landing location, such as the dry lakes at Edwards Air Force Base (AFB), to use the Rogallo wing.[67]

The preselected landing location had to lie within the downrange and cross-range footprint of the vehicle—since it was unpowered, the landing site had to be within the range of the vehicle's available energy—which was approximately 300 miles downrange and 50 miles across laterally.[68] An onboard digital computer was responsible for selecting a lift orientation that guided the spacecraft

67. Herbert G. Patterson et al., "Six-Degrees-of-Freedom Gemini Reentry Simulation," NASA Project Gemini Working Paper No. 5003, September 12, 1963, pp. 1–2.

68. David M. Box, Neil A. Armstrong et al., "Controlled Reentry," paper in the *Gemini Summary Conference*, Manned Spacecraft Center, Houston, TX, February 1–2, 1967, pp. 159–160.

to the landing location. Alternately, the astronauts could take manual control using the hand controller in the cabin. Offsetting the center of gravity approximately 1.75 inches from the longitudinal centerline resulted in an L/D of approximately 0.22.[69]

After initial development problems on the early Gemini flights, the guidance and control system worked well in both manual and automatic modes. On one occasion, the landing was made within a few hundred yards of the targeted point. The first use of a blunt-body lifting reentry vehicle verified that the concept would work for Apollo.[70]

The operational Gemini heat shield consisted of a tapered layer of McDonnell S-3 silicon elastomeric ablator 1 inch thick at the windward edge and 0.85 inch thick at the leeward edge (the heat shields for Gemini I and II were about half these thicknesses). The back structure was a 0.75-inch-thick section of phenolic fiberglass honeycomb consisting of two five-ply faceplates of resin-impregnated glass cloth separated by a 0.65-inch-thick fiberglass honeycomb core. The space between the heat shield and the back structure was filled with 0.375 inch of RF-300 batt insulation. The maximum diameter of the heat shield was 90 inches, 15.5 inches greater than Mercury's heat shield, and the spherical radius was 144 inches (resulting in a 1.6 ratio, instead of the 1.5 used on Mercury). The heat shield was attached to the cabin section by eighteen 0.25-inch-diameter bolts. The first production heat shield was completed on February 3, 1964.[71]

Interestingly, the S-3 ablator, developed under NASA contract, proved so effective that Dow Corning purchased the rights to the product and sold it as DC-325. The ablator formed an excellent char layer during ablation, was stable in a vacuum, and could withstand the wide variety of temperatures in the space environment. Perhaps the most important change was that McDonnell had figured out how to lay-up the cloth across the entire heat shield, eliminating the center plug used on Mercury. An MX-2625 Fiberite ring was used around the outer edge of the heat shield where extra strength was required to absorb loads transferred from the launch vehicle. This was a significant improvement

---

69. Patterson et al., "Six-Degrees-of-Freedom Gemini Reentry Simulation," pp. 1–2; William J. Blatz, "Gemini Design Features," copy in possession of authors.

70. Willis B. Mitchell et al., "Gemini Results as Related to the Apollo Program," paper in the *Gemini Summary Conference*, Manned Spacecraft Center, Houston, TX, February 1–2, 1967, pp. 329–330.

71. S.A. Mezines et al., "Gemini Heat Shield Trajectory Study," Thermodynamic Technical Note No. 33 (May 1963), no page numbers; James M. Grimwood and Barton C. Hacker, *Project Gemini Chronology: Technology and Operations* (Washington DC: NASA, 1969), p. 132; *Project Gemini Familiarization Manual*, July 18, 1963, pp. 2–12; Malik and Souris, "Project Gemini," pp. 13–14.

over the phenolic resin-impregnated laminated fiberglass unit used on Mercury. The Gemini heat shield weighed 347 pounds, compared to the Mercury heat shield's 303 pounds, mostly because of a 48-percent increase in area, a 25-percent increase in the ballistic loading coefficient, and a 90-percent increase in the design heat load (per square foot) due to the lifting reentry.[72]

The afterbody heat protection used on Gemini was almost identical to that used on Mercury. Beaded (corrugated) 0.016-inch-thick René 41 shingles were used over the conical section, and the outer surfaces of the shingles were coated with a black oxide film to enhance the thermal balance and protect the René 41 from oxidation. These shingles could withstand up to 1,800 °F by achieving a thermal balance through radiation to the atmosphere. The shingles were attached to the underlying structure using bolts through oversized holes that accommodated thermal expansion. Small blocks of Min-K insulation were used at the boltholes, and a layer of Thermoflex batt insulation was used between the shingles and the underlying structure. The cylindrical sections of the afterbody experienced temperatures too high for effective radiative cooling and used eight beryllium shingles as heat sinks. These shingles had a thin gold coating on the inside to minimize thermal radiation into the spacecraft and were 0.24 inch thick on the windward side and 0.09 inch thick on the leeward side.[73]

The first Gemini launch to use a heat shield was the Gemini 2 mission, which contained production units of all the equipment used on the later piloted missions except for the rendezvous radar, drogue parachute system, and auxiliary tape memory. The spacecraft was controlled through an automated sequencing device. Gemini 2 had been scheduled for launch on December 9, 1964, but as the booster engines were ignited, the Malfunction Detection System detected a loss of hydraulic pressure and shut down the engines 1 second later. After correcting the problems and checking the vehicle, Gemini 2 was launched on January 19, 1965. The capsule splashed down 2,125 miles downrange and was recovered by the USS *Lake Champlain* (CVS-39). During reentry, temperatures near the adapter interconnect fairing on the afterbody were higher than expected and damaged two René 41 shingles. Subsequent Gemini capsules were built with slightly heavier gauge shingles, and the trim angle of attack was lowered to reduce local heating.[74]

---

72. Blatz, "Gemini Design Features."

73. Ibid.; "Project Gemini Familiarization Manual," July 18, 1963, p. 2-7.

74. E. Rood and R. Posgay, "Thermodynamic Report for the Model 195 Heat Shield Qualification Spacecraft," McDonnell report E709 (July 15, 1966), p. 2-1; Malik and Souris, "Project Gemini," pp. 14, 36–37.

The Gemini 3 spacecraft, with flotation collar still attached, being hoisted aboard the USS *Intrepid* (CVS-11) during recovery operations following the successful Gemini-Titan 3 flight of March 1965. NASA S65-18656.

As far as is known, only a single capsule has been flown twice. The Gemini 2 capsule was refurbished and launched again on November 3, 1966, in a test of the Gemini-B[75] configuration for the Air Force's MOL program. This time, the vehicle rode atop a simulated MOL on a Titan IIIC from LC-40 at Cape Canaveral. The Gemini-B heat shield was identical to the standard Gemini heat shield with one exception: it had a 26-inch-diameter crew-transfer hatch cut into it to allow the astronauts to move between the capsule and laboratory. For the heat shield qualification (HSQ) flight, the RF-300 insulation used between the NASA heat shield and the capsule structure was not included since the reentry was expected to be more benign than the standard Gemini mission. In fact, the backside of the heat shield reached only 133 °F on the HSQ flight, compared to the Gemini's 430 °F design limit. Researchers were particularly interested to learn if the new hatch would adversely affect the ablative performance of the heat shield. The recovered capsule showed the performance was satisfactory.[76]

The operational experience with the Gemini heat shield was decidedly better than Mercury, mostly because the troublesome center plug had been eliminated. The first piloted Gemini mission, the GT-3, piloted by Gus Grissom and John W. Young on March 23, 1965, showed that the heating effect on the heat shield appeared much less severe than was noted on Gemini 2. After the heat shield was removed from the spacecraft and dried, engineers noted it had lost about 20 pounds due to ablation, which was about average for the subsequent missions.[77]

Only two Gemini missions suffered any notable damage to their thermal protection systems. After Gemini VII, piloted by Frank Borman and James A. Lovell, Jr., on December 4, 1965, engineers found a small hole in one René 41 shingle, just forward of the right-hand hatch, and a beryllium shingle around the bottom of the C-band antenna shifted its position enough to partially

---

75. Gemini-B was different from Blue Gemini. Gemini-B were purpose-built capsules intended for the MOL program and included the crew transfer hatch in the heat shield and various other improvements to allow the capsule to remain in orbit for prolonged periods docked to the MOL. Blue Gemini was an unrealized plan to fly Air Force astronauts using standard NASA Gemini capsules.

76. Rood and Posgay, "Thermodynamic Report for the Model 195 Heat Shield Qualification Spacecraft," pp. 2-1, 3-1.

77. "Gemini Program Mission Report, Gemini 3," NASA MSC-G-R-65-2 (April 1965), pp. 1-1, 1-2, 12-5.

As a helicopter flies above, the Gemini 12 spacecraft with parachute open descends to the Atlantic with astronauts Jim Lovell and Edwin E. "Buzz" Aldrin aboard. NASA 66C-9516.

cover the antenna. Neither flaw was considered critical.[78] On the last mission of the program, Gemini XII, with Lovell and Edwin E. "Buzz" Aldrin, Jr., on November 11, 1966, the postflight inspection revealed that the left-hand skid door had a shingle curled on the corner and a washer missing, and the forward lower centerline equipment door was deflected inward. The damage to both was subsequently determined to be caused by landing impact, but none of the anomalies was considered critical.[79]

Project Apollo

Although it was the third piloted U.S. program to fly, several Apollo studies predated even Mercury. Because of this timing, the Apollo spacecraft was not an evolution of Mercury or Gemini, and its development took place largely uncoupled from the smaller programs. Surprisingly to many people, Apollo was an approved program before President John F. Kennedy made his now famous speech challenging the Nation to land an American on the Moon before the end of the 1960s. At the time, a lunar landing was not planned, and the spacecraft was supposed to be capable of piloted circumlunar flight, along with several Earth-orbital tasks.[80]

Various reentry configurations were investigated for the Apollo spacecraft, including capsules and lifting reentry vehicles using lenticular shapes, lifting bodies, and wings. The lifting reentry shapes were considered desirable since they offered the capability of a horizontal landing in a relatively conventional manner on a prepared surface. However, these designs presented several drawbacks, chiefly that they generally weighed several thousand pounds more than capsules, and that they were able to cope with only prepared landing surfaces. It was also noted that the cross range provided by the lifting vehicles was not required by any of the mission models. Before the issue could be decided on technical merits, President Kennedy committed the Nation to a space race, and

---

78. "Gemini Program Mission Report, Gemini VII," NASA MSC-G-R-66-1 (January 1966), pp. 1-1, 1-2, 12-27. The postflight reports do not include a weight for the heat shield.

79. "Gemini Program Mission Report, Gemini XII," NASA MSC-G-R-67-1 (January 1967), pp. 1-1, 1-2, 12-38.

80. Robert O. Piland, Caldwell J. Johnson, Jr., and Owen E. Maynard, "Apollo Spacecraft Design," *NASA-Industry Apollo Technical Conference: A Compilation of the Papers Presented* (July 18–20, 1961), p. 1.

researchers decided a capsule was the most expedient course. The spaceplane would have to wait.[81]

During the late 1950s and into the 1960s, the three major human space programs—Mercury, Dyna-Soar, and Apollo—faced very different reentry concerns. Mercury would perform a ballistic reentry from a low-Earth orbit (or a ballistic trajectory for the suborbital flights). The heating rates and durations it faced were much the same as the ballistic missiles, and the total size of the capsule was not much larger or heavier. Thus, the solutions used to protect the missile warheads were generally adequate for the Mercury capsule. Dyna-Soar was a winged vehicle that used a lifting reentry, which presented a lower heating rate than did ballistic reentry but for a much longer duration. The desire to use the glider for multiple missions dictated a metallic heat shield almost by default. Apollo needed to withstand superorbital entries after its lunar flybys, resulting in a heating load an order of magnitude greater than Mercury, the ballistic missiles, or Dyna-Soar. Because Apollo would fly a semilifting entry, the heating duration was also quite long, although it was not as extreme as that of Dyna-Soar.

Engineers at NASA and prime contractor North American Aviation estimated that the total heat at the stagnation point might be as high as 120,000 BTU per square foot, with stagnation point heating rates of 520 BTU per square foot per second.[82] It was clear that the only material that could withstand the heating at the stagnation point was an advanced ablator. The best material for the rest of the heat shield, and indeed the rest of the capsule, was less clear. For instance, Faget, one of the principle designers of Mercury, wanted to use René 41 and beryllium shingles on the afterbody, as had been done with Mercury, but the heating estimates were higher than could be tolerated by a shingle of any practicable thickness.[83]

In 1961, the possible heat shield materials that could withstand the Apollo environmental conditions included ablators, graphites, ceramics, and metals. Since the expected stagnation point heating rate was about 600 BTU per square foot per second, it was clear that ablators were required near the stagnation point. However, as the heating rate tapered off around the body, there

---

81. Piland, Johnson, and Maynard, "Apollo Spacecraft Design," pp. 7–8.

82. "Apollo Final Report," Martin Report ER-12010-2, Submitted to the Space Task Group, Contract NAS5-303 (June 1961), p. I-1.

83. Courtney G. Brooks, James M. Grimwood, and Loyd S. Swenson, Jr., *Chariots for Apollo: A History of Manned Lunar Spacecraft* (Washington, DC: NASA, 1979), p. 37.

were locations where various metal alloys, resin-impregnated ceramics, or plain ceramics might be used.[84]

Because Apollo would use a lifting reentry, followed by a precision landing on land, the aerodynamic qualities of the capsule were important; the outer mold line needed to be maintained during entry. This made the concept of a metallic heat shield attractive, and various NASA and contractor researchers looked into superalloy and refractory metal heat shields similar to those being studied for Dyna-Soar. The materials included cobalt-base Haynes 25 and nickel-base René 41 superalloys. However, to withstand the superorbital entry conditions, either superalloy would have to be used as part of a Bell-type double-wall structure, most likely with an active water-cooling system, resulting in a very heavy installation. It was possible that the refractory metal, such as titanium-molybdenum alloy and F-48 columbium alloy, could withstand the heating conditions without melting, but the metals suffered severe oxidation during entry, and no suitable coatings had yet been developed. Without these coatings, the refractory metals could not withstand the expected environment without the risk of structural failure.

Efforts within the Martin Company resulted in the development of highly porous ceramics, including foamed zirconium oxide, foamed silicon carbide coated with zirconium oxide, foamed aluminum oxide, and foamed vitreous silicon dioxide. These materials exhibited temperature limits of 4,000 °F, 3,300 °F, and 2,600 °F, respectively. The major advantage of the ceramic foams was that they retained their shape better than any ablator available in 1961. Unfortunately, they weighed considerably more than the equivalent ablators.[85]

The consensus of NASA researchers at Langley was that the entire capsule should be coated in ablator, which would also provide a small amount of radiation protection during translunar flight.[86] However, the aerodynamic concerns still existed, and the retention of the char layer was critical to maintain the outer mold line of the vehicle and preserve the lifting entry characteristics. However, initial studies into the nylon-phenolic ablator that most nearly met these requirements showed that its char was of low strength and would likely spall due to aerodynamic pressures. This led to studies of fibrous additions to strengthen the char and ensure the retention of thick char layers. Eventually, Avco (now part of Textron) developed a superior nonreceding ablator named "Avcoat." The development was largely based on research into materials for the high-ballistic-coefficient long-range missile warheads.

---

84. "Apollo Final Report," p. I-1.

85. Ibid., p. V-1.

86. Brooks, Grimwood, and Swenson, *Chariots for Apollo*, p. 37.

The wind tunnel experts at Langley investigated heat transfer, heating loads and rates, and aerodynamic and hydrodynamic stability on the Command Module heat shield. Avco originally proposed a layered and bonded heat shield similar to the one General Electric made for Mercury. In the interim, however, McDonnell advanced heat protection technology by developing ablator-filled honeycomb material for Gemini. This appeared to be a significant advancement, and Avco refined the new system to withstand the higher heating rates of superorbital entry. For the Gemini heat shield, a fiberglass honeycomb material was bonded to the back-structure and the S-3 ablator was poured into it and allowed to harden.[87] Because of the viscosity of the Avcoat ablator, this technique would not work.

For Apollo, a brazed PH 14-8 stainless steel honeycomb sheet was attached to the structural shell, and a fiberglass-phenolic honeycomb with 400,000 individual cells was bonded to it. Then, each cell was individually filled with a mixture of silica fibers and micro-balloons using a caulking gun, making the manufacturing process very labor intensive.[88] The pot life of the Avcoat-5026 was short, and quantities of it were frequently thrown away because technicians could not fill the cells quickly enough before it began to cure. It was a tedious process. The finished heat shield weighed about 32 pounds per cubic foot. The structural shell was attached to the aft bulkhead of the Command Module using 59 bolts installed in oversize holes to permit the heat shield to move relative to the inner structure.[89]

Given the limitation of the available ground test facilities, researchers knew they would not completely understand the conditions associated with superorbital entry prior to the first lunar mission. Nevertheless, they instituted Project FIRE (Flight Investigation of the Reentry Environment) to obtain data on materials, heating rates, and radio-signal attenuation on spacecraft reentering the atmosphere at speeds of up to 24,500 mph. NASA Langley announced the program, which involved both wind tunnel and flight tests, on February 18, 1962. The wind tunnel tests used several Langley facilities, including the Unitary Plan Wind Tunnel, the 8-foot High Temperature Tunnel, and the 9-by-6-foot Thermal Structures Tunnel.[90]

---

87. Ibid., p. 92.

88. "NASA's Exploration Systems Architecture Study," NASA TM-2005-214062 (November 2005), pp. 290–291.

89. P.D. Smith, "Apollo Experience Report: Spacecraft Structure Subsystem," NASA TN-D-7780 (October 1974).

90. Brooks, Grimwood, and Swenson, *Chariots for Apollo*.

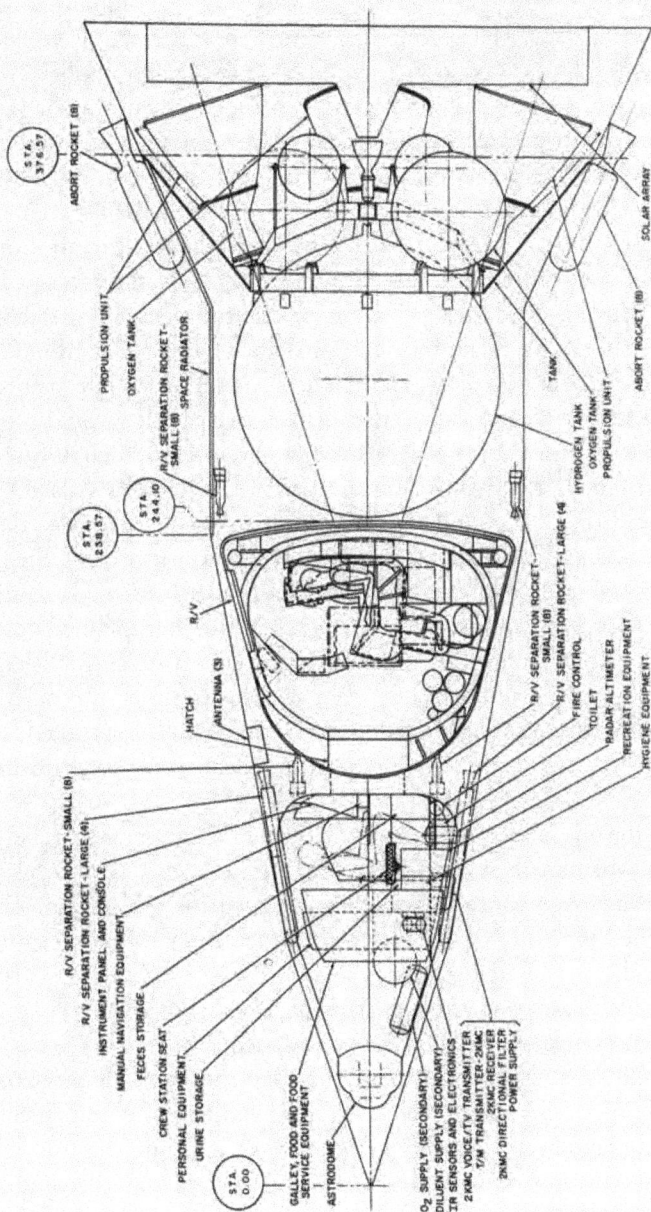

Figure III-2-2. Recommended APOLLO Configuration

STA. 376.57

STA. 258.57

STA. 244.10

STA. 0.00

ABORT ROCKET (8)

PROPULSION UNIT

OXYGEN TANK

R/V SEPARATION ROCKET-SMALL (8) SPACE RADIATOR

R/V

HATCH

ANTENNA (3)

R/V SEPARATION ROCKET-SMALL (8)

R/V SEPARATION ROCKET-LARGE (4)

INSTRUMENT PANEL AND CONSOLE

MANUAL NAVIGATION EQUIPMENT

FECES STORAGE

CREW STATION SEAT

PERSONAL EQUIPMENT,
URINE STORAGE

GALLEY, FOOD AND FOOD
SERVICE EQUIPMENT

ASTRODOME

O₂ SUPPLY (SECONDARY)
DILUENT SUPPLY (SECONDARY)
IR SENSORS AND ELECTRONICS
2KMC VOICE/TV TRANSMITTER
T/M TRANSMITTER-2KMC
2KMC RECEIVER
2KMC DIRECTIONAL FILTER
POWER SUPPLY

SOLAR ARRAY

ABORT ROCKET (8)

TANK

HYDROGEN TANK
OXYGEN TANK
PROPULSION UNIT

R/V SEPARATION ROCKET-
SMALL (8)

R/V SEPARATION ROCKET-LARGE (4)

FIRE CONTROL

TOILET

RADAR ALTIMETER

RECREATION EQUIPMENT

HYGIENE EQUIPMENT

III-4

This 1961 General Electric concept for the Apollo spacecraft has a more X-15-like look to it than the blunt-body capsule that was actually built. NASA.

On March 29, 1962, NASA Langley awarded a $5 million contract to Republic Aviation for two experimental reentry spacecraft that were essentially subscale models of the Apollo Command Module. On November 27, 1962, NASA awarded a $2.56 million contract to Ling-Temco-Vought (LTV) to develop the velocity package that would simulate reentry from a lunar mission. The velocity package was essentially an Antares II A5 (ABL X-259d) solid-propellant rocket motor manufactured by the Allegheny Ballistic Laboratory. A heat shroud, manufactured for LTV by the Douglas Aircraft Company, protected the spacecraft from aerodynamic heating during the boost ascent. The velocity package also included a guidance system for maintaining stability and control, a telemetry system for transmitting flight data, and an ignition/destruct system.[91] An Atlas-D booster would lift the Republic spacecraft to an altitude of 400,000 feet (entry interface), then the velocity package would propel the reentry vehicle into a –15-degree trajectory at a velocity of 37,000 ft/sec.[92]

The first Project FIRE flight was on April 14, 1964. The Atlas-D launch vehicle placed the 200-pound spacecraft into a ballistic trajectory along the Eastern Test Range. The Antares fired for 30 seconds, increasing the descent speed to 25,166 mph, and the exterior of the spacecraft reached an estimated 20,000 °F. About 32 minutes after launch, the spacecraft splashed down into the Atlantic Ocean some 5,000 miles downrange, near Ascension Island. The second Project FIRE launch, following a similar trajectory, was on May 22, 1965, and also used an Atlas-D from Cape Canaveral. The spacecraft entered the atmosphere at a speed of 25,400 mph and generated temperatures of about 20,000 °F.[93]

Despite the information provided by Project FIRE, researchers believed that large unknowns remained concerning superorbital reentry, so NASA, North American, and Avco conservatively overdesigned the heat shield. The heat shield weighed almost 1,500 pounds, over 10 percent of the entire command-module weight. In fact, none of the lunar-return missions used even 20 percent of the available ablator, and engineers later decided they could have cut the heat shield thickness in half and still have had an adequate margin.

The AS-201 spacecraft/launcher test of a Saturn IB on February 26, 1966, had several objectives, one of which involved verifying the heat shield's adequacy during high heat-rate reentry from low-Earth orbit. This was only partially achieved because of a fault in the electrical power system, causing a loss of

---

91. *http://www.voughtaircraft.com/heritage/products/html/fire.html*, accessed August 23, 2009.

92. Brooks, Grimwood, and Swenson, *Chariots for Apollo*.

93. W.C. Kuby et al., "Analysis of the Project Fire Reentry Package Flow Field," Philco Corp. Report U-3020 (October 8, 1964).

High-angle view of Command Module 012, looking toward the –Z axis, during preparation for installation of the crew compartment heat shield—showing technicians working on aft bay. This was the ill-fated Apollo 1 spacecraft. NASA s66-41851.

steering control that resulted in a rolling reentry. This meant that data on the heat shield's effectiveness was insufficient to affirm confidence in the system. Even so, as Apollo's official historians concluded: "AS-201 proved that the spacecraft was structurally sound and, most important, that the heat shield could survive an atmospheric reentry."[94] A second test of the Apollo spacecraft/ Saturn IB, AS-202, was both more complex and more successful than the earlier one. Launched on August 25, 1966, this flight was intended to demonstrate the structural integrity and compatibility of the launch vehicle, the Command and Service Modules, and a host of other objectives, including the adequacy of

94. Brooks, Grimwood, and Swenson, *Chariots for Apollo*, p. 193; "Postlaunch Report for Mission AS-201 (Apollo Spacecraft 009)," NASA-TM-X-72334 (May 6, 1966), pp. 5-1, 5-2.

the heat shield for reentry and the quality of the ablator during high reentry temperatures and pressures. The mission was a success, and NASA declared that all objectives had been achieved. In terms of the reentry sequence, the heat shield and all other components performed acceptably despite a steeper reentry than had been intended. This meant that the spacecraft landed 205 nautical miles uprange from the planned point and experienced greater g-forces than expected. Even so, heat shield engineers found that all components, in the words of NASA's Apollo historians, "had come through reentry admirably." As the postlaunch report noted: "During entry, spacecraft attitude was controlled to provide a skip trajectory resulting in a double-peak heating-rate history."[95]

These tests made NASA engineers confident that they could proceed with the first Apollo capsule flight tests while undertaking test flights of the Saturn V launch vehicle, testing the capsules and their thermal protection systems under high-speed reentry conditions. Apollo 4 tested the worst-case scenario shallow reentry, which was long and drawn out with a maximum total heat load; and Apollo 6 tested the worst-case scenario steep reentry, which was short and sharp with maximum heating rate. The Apollo 4 mission was launched on November 9, 1967, marking the first use of a Saturn V. The payload consisted of a Block I Command/Service Module (spacecraft 017) and a Lunar Module test article (LTA-10R). Postflight inspection revealed that the Block II thermal protection system survived the lunar entry environment satisfactorily, and the aft ablative heat shield was heavily charred over its entire surface. Temperature data indicated surface temperatures had exceeded 5,000 °F, but the maximum char penetration was 0.88 inch, compared to the 1.25 inches expected.[96]

Apollo 6 was launched on April 6, 1968, with an unpiloted Block I Command/Service Module (spacecraft 020) and a Lunar Module test article (LTA-2R). Despite numerous problems during ascent, the mission was generally successful. Oddly, although Apollo 6 entered at 3,600 ft/sec slower than Apollo 4 had, the temperature on the conical section and on the leeward side of the heat shield was substantially higher than that of the earlier mission. This paradoxical situation can be attributed to three causes: Apollo 6 flew faster at lower altitudes, it did not skip out to as great an altitude to allow the ablator to cool down, and it flew approximately 80 seconds longer to reach the desired splashdown point. Despite the higher than expected temperatures,

95. Brooks, Grimwood, and Swenson, *Chariots for Apollo*, p. 193; "Postlaunch Report for Mission AS-202 (Apollo Spacecraft 011)," MSC-A-R-66-5 (October 12, 1966), p. 1-2.

96. "Apollo 4 Mission Report," NASA Report MSC-PA-R-68-1 (January 7, 1968), p. 1-1, 5.4-7, 5.4-8. The Apollo 5 mission tested the Lunar Module in orbit and did not carry a Command Module.

The Apollo 11 Command Module (107) is loaded aboard a Super Guppy aircraft at Ellington AFB for shipment to the North American Rockwell Corporation in Downey, CA. The Command Module was just released from its postflight quarantine at the Manned Spacecraft Center. The Apollo 11 spacecraft was flown by astronauts Neil A. Armstrong, commander; Michael Collins, Command Module pilot; and Edwin E. Aldrin, Jr., Lunar Module pilot, during their lunar landing mission. Note damage to aft heat shield caused by extreme heat of Earth reentry. NASA s69-41985.

postflight inspection showed that the heat shield performed satisfactorily. During the inspection, however, engineers noticed that ablator was missing around some of the manufacturing splices in the heat shield. Since the edges around the missing material were not burned, the material must have separated after entry. Engineers had noted a similar anomaly on the AS-202 Command Module but not on the Apollo 4 capsule. The common thread was that AS-202 and Apollo 6 had remained in the water for much longer than Apollo 4 had. Improved manufacturing processes eliminated the splices between the honeycomb segments on all Block II heat shields used for the lunar missions.[97]

---

97. "Apollo 6 Mission Report," NASA report MSC-PA-R-68-9 (May 1968), pp. 1-1, 5.4-1 through 5.4-3, 5.4-13.

Apollo 11 Command Module heat shield inspected on November 6, 1967. NASA S67-48970.

The operational experience with the Apollo heat shield was highly satisfactory, with no significant anomalies noted on any of the missions. Apollo 8, the first lunar flyby mission in December 1968, returned with a heat shield that was charred less than that of the Apollo 4 test. The char depth was 0.6 inch at the stagnation point, and only 0.4 inch elsewhere. There was, however, a fair amount of impact damage when the capsule splashed down. Apollo 10, another lunar return, entered at a flightpath angle 0.02 of a degree steeper than planned, subjecting the crew to a maximum 6.78 g's. Despite this anomaly, the heat shield performed satisfactorily. Returning from the first lunar landing, Apollo 11 crossed the entry interface at 400,000 feet and at 36,195 ft/sec at a flightpath angle of –6.488 degrees (versus a planned 36,194 ft/sec and –6.483 degrees). Heat shield performance was listed as nominal. Despite Apollo 13's faster than normal entry, Lovell reported that the craft's landing decelerations were mild in comparison to Apollo 8. The postflight report revealed nothing unexpected about the performance of the heat shield. The thermal protection system worked as expected on Apollo 15, but there was an anomaly regarding the recovery parachutes. At approximately 10,500 feet altitude, the three main parachutes deployed normally, but 4,000 feet later, one collapsed and deflated. Engineers believed that an unexpected dump of monomethyl hydrazine caused two aluminum riser connectors to fail, and these parts were changed to Inconel 718 for future flights. Although the capsule impacted the Pacific Ocean at a higher-than-normal velocity, the capsule was not damaged nor was the crew injured. The remaining lunar-return flights were accomplished without incident regarding the heat shield or parachute performance.[98]

---

98. "Apollo 7 Mission Report," NASA Report MSC-PA-R-68-15 (December 1968), pp. 1-1, 5-24 through 5-30; "Apollo 8 Mission Report," NASA Report MSC-PA-R-69-1 (February 1969), pp. 1-1, 6-13 through 6-14; "Apollo 9 Mission Report," NASA Report MSC-PA-R-69-2 (May 1969), pp. 1-1, 8-1; "Apollo 10 Mission Report," NASA Report MSC-00126 (August 1969), pp. 1-1, 6-5; "Apollo 11 Mission Report," NASA Report MSC-00171 (November 1969), pp. 1-1, 8-8 through 8-9; "Apollo 12 Mission Report," NASA Report MSC-01855 (March 1970), p. 1-1; "Apollo 13 Mission Report," NASA Report MSC-02680 (September 1970), pp. 1-1, 8-17; "Apollo 14 Mission Report," NASA Report MSC-04112 (May 1971), p. 1-1; "Apollo 15 Mission Report," NASA report MSC-05161 (December 1971), pp. 1-1, 6-1, 14-17 through 14-20; "Apollo 16 Mission Report," NASA Report MSC-07230 (August 1972), p. 1-1; "Apollo 17 Mission Report," NASA Report JSC-07904 (March 1973), p. 1-1.

Project Mercury recovery practice aboard USS *Strong* on August 17, 1959. NASA.

CHAPTER 3

Landing Under Canopies

Returning from space is just as difficult and risky as reaching into space. Indeed, coming home may well be just as risky as launch. For the United States, NASA has undertaken 168 flights (counting suborbital), including 135 for the Space Shuttle program. In addition, Russia (or the Soviet Union) has undertaken 119 flights, mostly for the Soyuz spacecraft program (including suborbital flights); China has made 3 flights; and the private SpaceShipOne completed 3 flights (suborbital flights) for a total of 293 human space flights throughout the more than 50-year-long space age.[1] During this time, there have been four accidents resulting in the loss of crew for a 1.36-percent failure rate, by flight. These accidents include the following:

- April 24, 1967: A parachute failure caused the Soyuz spacecraft to hit the ground at 300 mph, killing one cosmonaut.
- June 30, 1971: Three cosmonauts died on descent when their spacecraft decompressed prematurely.
- January 28, 1986: On ascent, the Space Shuttle Challenger broke apart, killing seven crewmembers.
- February 1, 2003: On descent, the Space Shuttle Columbia broke apart, killing seven crewmembers.

Significantly, the number of failures on descent exceeds the number of ascent failures, even though many treat ascent as the more risky venture. Accordingly, the challenge of returning to Earth is a subject of great significance. How might it most expeditiously be accomplished for the human spaceflight program?

Virtually all of the early concepts for human space flight involved spaceplanes that flew on wings to runway landings. Eugen Sänger's antipodal bomber of the 1920s did so, and Wernher von Braun's popular concepts of the 1950s did the same. However, these concepts proved impractical for launch vehicles

---

1. These include 2 suborbital X-15 flights, 2 suborbital Mercury flights, 4 orbital Mercury flights, 10 Gemini flights, 12 Apollo flights, 3 Skylab flights, and 135 Space Shuttle flights; 6 Vostok flights, 2 Voskhod flights, 1 suborbital Soyuz flights, 110 orbital Soyuz flights, 3 Shenzhou flights, and 3 suborbital SpaceShipOne flights.

available during the 1950s, and capsule concepts that returned to Earth via parachute proliferated, largely because they represented the art of the possible at the time. The United States was racing against the Soviet Union for primacy in space, which necessitated placing a human in space as expeditiously as possible. In such a Cold War environment, pursuing what was possible rather than what was desirable proved irresistible.

All of the U.S. spacecraft up to the Space Shuttle, as well as the Soviet/Russian and Chinese piloted capsules, used from one to three parachutes for return to Earth. The American capsules landed in the ocean and were recovered by ship. This exposed them to corrosive saltwater while waiting for expensive recovery efforts. The Soviet/Russian and Chinese spacecraft have always been recovered on land, which presented the crew with a harder landing than would be the case at sea. For Project Gemini, NASA toyed with the possibility of using a paraglider that the Langley Research Center was developing for "dry" landings, instead of "splashdowns" in water and recovery by the Navy. At sum, this represented an attempt to transform the capsule into a spaceplane. Unfortunately, the engineers never did get the paraglider to work properly and eventually dropped it from the program in favor of a parachute system like the one used for Mercury.

During the period between the creation of NASA in 1958 and the Moon landings of the late 1960s and early 1970s, NASA developed capsules with blunt-body ablative heat shields and recovery systems that relied on parachutes. Most people are familiar with parachutes, the umbrellalike devices usually made of a soft fabric that slow the motion and retard the descent of a falling object by creating drag as it passes through the air. The parachute's seeming simplicity masked significant challenges for recovery from space as NASA pondered the details of how to implement it successfully for landings from space. Every piloted U.S. mission that has used parachutes has been successful, having safely returned crews from space back to Earth via water landings. However, history did not have to unfold in that way.

The Department of Defense (DOD) tested parachute-landing systems during Project SCORE in 1958 and employed the concept throughout the CORONA satellite reconnaissance program of the 1960s, when it would snatch in midair return capsules containing unprocessed surveillance footage dangling beneath parachutes. During the Mercury program, astronauts rode a blunt-body capsule with an ablative heat shield to a water landing and rescue at sea by the Navy. Gemini later used a similar approach, but NASA engineers experimented with a Rogallo wing and a proposed landing at the Flight Research Center on skids similar to those employed on the X-15. When the engineers working on the Rogallo wing concept failed to make the rapid progress required to meet project timelines, NASA returned to the parachute concept used in Project Mercury. Engineers incorporated essentially the same approach used in the Apollo program, although with greatly improved ablative heat shields and much larger and more complex parachute systems.

Pioneering Recovery: Project Mercury

In theory, the deployment of a parachute for a gentle landing upon returning to Earth was the most simple of engineering tasks. Such was not to be the case, however, because of a succession of challenges ranging from hypersonic reentry to alternative freezing and overheating of the system while in Earth's orbit. As intended during Mercury, the spacecraft would undergo reentry and, after deceleration to about 270 mph, "at about 21,000 feet a six-foot diameter drogue parachute [would] open to stabilize the craft. At about 10,000 feet, a 63-foot main landing parachute [would] unfurl from the neck of the craft. On touchdown, the main parachute [would] jettison."[2] A landing bag, inflated from behind the heat shield, would deploy to soften the impact just prior to hitting the water. Upon landing, additional bags inflated around the nose of the craft to keep the capsule upright in the water, and the parachutes were released. Finally, the astronaut would open the hatch only after Navy frogmen had secured the capsule and a recovery helicopter had connected to the vehicle.[3]

The full-scale testing of the Mercury recovery system began at the Langley Research Center, Hampton, VA, on October 9, 1958. The first task was to undertake tests of parachute deployment and spacecraft stability, first from helicopters and later from U.S. Air Force C-130s at Pope AFB, NC, where the 317th Military Airlift Wing was located. These tests were relatively simple, at first using a concrete-filled 55-gallon drum attached to a deployment system. As the official NASA history of Project Mercury concluded, "By early January more than a hundred drops of drums filled with concrete and of model capsules had produced a sizable amount of evidence regarding spacecraft motion in free falls, spiraling and tumbling downward, with and without canopied brakes, to impacts on both sea and land."[4] These efforts demonstrated the adequacy of the

---

2. "Mercury-Atlas 6 at a Glance," NASA release 62-8, NASA Historical Reference Collection, p. 1–6.

3. "Mercury Parachute," *Astronautix*, available online at *http://www.astronautix.com/craft/merchute.htm*, accessed August 27, 2009.

4. "Status Report on Project Mercury Development Program as of March 1, 1959," Public Affairs Office, Langley Research Center, as cited in Swenson, Grimwood, and Alexander, *This New Ocean*, p. 141.

The sequence of events, from launch to parachute opening, of a beach abort test for the Mercury capsule with a launch escape motor on April 13, 1959. NASA L59-2768.

mechanical system proposed to deploy parachutes from the Mercury spacecraft. Using the C-130s, NASA graduated to using full-scale spacecraft and operating parachutes on these drop tests; in the process, NASA was working out the operational features of the parachute system and the processes necessary to ensure success in returning from space. The project then moved to Wallops Island, VA, and continued tests, "to study the stability of the spacecraft during free fall and with parachute support; to study the shock input to the spacecraft by parachute deployment; and to study and develop retrieving operations."[5]

Not everything went as planned in these tests. NASA engineers found that the main parachute was prone to experience "squidding," a phenomenon also referred to as "breathing" or "rebound." Because of conditions in the atmosphere between an approximate range of 70,000 feet and 10,000 feet, there were "snatch" forces, shock waves, and stability difficulties that resulted in the partial opening of the parachute—occurrences and conditions that proved unacceptable for the spacecraft returning to Earth. These problems were corrected through the acquisition of a 63-foot-diameter ribbon ring-sail parachute. This ribbon ring-sail parachute, developed by Theodor W.

---

5. "Mercury Parachute," *Astronautix*, available online at *http://www.astronautix.com/craft/merchute.htm*, accessed August 27, 2009.

Knacke, had superior stability than earlier parachute designs, making it ideal for human space programs.[6] The parachute change resolved the problems that arose during testing and served effectively throughout the Mercury, Gemini, and Apollo programs.

For flight tests, NASA acquired 6 main-parachute and 12 drogue-parachute canisters from the Goodyear Aircraft Corporation—3 each to be dedicated to the Little Joe and Big Joe tests of the Mercury capsule. These found use in a test effort concerned with the problem of landing impact, specifically the question of whether the spacecraft could touch down at a speed of less than 30 ft/sec. Ensuring this capability, as well as a vertical landing, proved an exasperating challenge. NASA initially hoped for a land recovery, but the shock of impact was too great. NASA then pursued landing tests in hydrodynamics laboratories. Reducing the speed prior to impact required such efforts as the following:

> McDonnell engineers fitted a series of four Yorkshire pigs into contour couches for impact landing tests of the crushable aluminum honeycomb energy-absorption system. These supine swine sustained acceleration peaks from 38 to 58 g before minor internal injuries were noted. The "pig drop" tests were quite impressive, both to McDonnell employees who left their desks and lathes to watch them and to STG engineers who studied the documentary movies. But, still more significant, seeing the pigs get up and walk away from their forced fall and stunning impact vastly increased the confidence of the newly chosen astronauts that they could do the same.[7]

As McDonnell engineers concluded, "Since neither the acceleration rates nor shock pulse amplitudes applied to the specimens resulted in permanent or disabling damage, the honeycomb energy absorption system of these experiments is considered suitable for controlling the landing shock applied to the Mercury capsule pilot."[8]

Notwithstanding these tests, cushioning the blow to astronauts during landing and recovery operations led to a succession of project innovations. The first was the creation of form-fitting contour couches that helped distribute the

---

6. George M. Low, "Status Report No. 11" (April 6, 1959); Swensen, Grimwood, and Alexander, *This New Ocean*, p. 143.

7. Swensen, Grimwood, and Alexander, *This New Ocean*, p. 144.

8. "Pilot Support System Development (Live Specimen Experiment)," Report 6875, McDonnell Aircraft Corp. (June 1959), NASA Historical Reference Collection.

g-forces associated with reentry and landing throughout the body, allowing astronauts to withstand the loads without undue injury. During flight, "the astronaut, wearing the Mercury full-pressure suit, was positioned in his contour couch in the semisupine position and secured by shoulder and lap harnesses." NASA also investigated the feasibility of including a crushable honeycomb-shape section of metal foil between the heat shield and the capsule to absorb the principle shocks of landing. In the end, NASA included a pneumatic landing bag in the Mercury capsule.[9]

In June 1959, NASA contracted with Northrop to design and fabricate the landing system for Project Mercury. Northrop had a long history in this arena, having built aircraft landing and recovery systems as far back as 1943, when it developed the first parachute-recovery system for pilotless aircraft. Northrop adopted the 63-foot ring-sail main parachute. By August 1959, Northrop had completed initial drop tests for the 63-foot ring-sail main parachute. It had also changed the drogue parachute configuration, taking it from 19.5 percent porosity with a flat circular ribbon chute to a 28-percent porosity, 30-degree conical canopy. The drogue parachute quickly moved through its certification and was qualified for deployment at speeds up to Mach 1.5 and altitudes up to 70,000 feet. This was far beyond the standard envelope for drogue chute operations: deployment at 40,000 feet and below Mach 1. By September 19, 1959, after approximately 18 months of effort, NASA completed the qualification tests for the Mercury spacecraft-landing system. In all, project staff and contractors had undertaken "56 airdrops of full-scale engineering models of the Mercury spacecraft from C-130 aircraft at various altitudes up to 30,000 feet and from helicopters at low altitudes to simulate off-the-pad abort conditions."[10]

Full-scale tests of the system in space began on December 4, 1959, when a mission flying Sam, an American-born Rhesus monkey, tried out the recovery system in a launch of Mercury Little Joe 2 (LJ-2) at Wallops Island, VA. Controllers initiated an abort sequence after 59 seconds of flight at an altitude of 96,000 feet and a speed of Mach 5.5. The drogue and main parachutes deployed properly and returned Sam safely to the ocean. Recovery of the capsule took about 2 hours, but Sam came through the flight fine. Additional tests thereafter found no difficulties with the system.[11]

---

9. *Results of the Second U.S. Manned Orbital Space Flight* (Washington, DC: NASA SP-6, May 24, 1962), p. 54.

10. "Mercury Parachute," *Astronautix*, available online at *http://www.astronautix.com/craft/merchute.htm*, accessed August 27, 2009.

11. Eugene M. Emme, *Aeronautics and Astronautics: An American Chronology of Science and Technology in the Exploration of Space, 1915–1960* (Washington, DC: NASA, 1961), p. 189ff.

Little Joe 1 (LJ-1) was launched on October 4, 1959, at Wallops Island, VA. This was the first attempt to launch an instrumented capsule with a Little Joe booster. Only the LJ-1A and the LJ-6 used the space metal/chevron plates as heat reflector shields, as they kept shattering. Little Joe was used to test various components of the Mercury spacecraft, such as the emergency escape rockets. NASA L-1960-00104.

Further tests of the Mercury capsule's extended-mission parachutes, under the name Project Reef, took place in 1962. Beginning on June 26, 1962, Project Reef began 20 airdrops to test the ability of the Mercury 63-foot ring-sail main parachute's capability to support a higher projected spacecraft weight for an extended-range or more than 1-day mission. Tests indicated that the parachute could support the heavier spacecraft without undue stress.[12]

For the flights of Project Mercury, the configuration of the parachute-recovery system consisted of the following elements:

> Above the astronaut's cabin, the cylindrical neck section contains the main and reserve parachute system.
>
> Three parachutes are installed in the spacecraft. The drogue chute has a six-foot diameter, conical, ribbon-type canopy with approximately six-foot long ribbon suspension lines, and a 30-foot long riser made of dacron to minimize elasticity effects during deployment of the drogue at an altitude of 21,000 feet. The drogue riser is permanently attached to the spacecraft antenna by a three point suspension system terminating at the antenna in three steel cables, which are insulated in areas exposed to heat.
>
> The drogue parachute is packed in a protective bag and stowed in the drogue mortar tube on top of a light-weight sabot or plug. The sabot functions as a free piston to eject the parachute pack when pressured from below by gasses generated by a pyrotechnic charge.
>
> The function of the drogue chute is to provide a backup stabilization device for the spacecraft in the event of failure of the Reaction Control and Stabilization System. Additionally, the drogue chute will serve to slow the spacecraft to approximately 250 feet per second at the 10,000 foot altitude of main parachute deployment.
>
> The reserve chute is identical to the main chute. It is deployed by a flat circular-type pilot chute.
>
> Other components of the landing system include drogue mortar and cartridge, barostats, antenna fairing ejector, and sea marker packet.
>
> Following escape tower separation in flight, the 21,000 and 10,000 foot barostats are armed. No further action occurs until the spacecraft descent causes the 21,000 foot barostat to close, activating the drogue ejection system.

---

12. NASA-MSC Report, "Project Mercury [Quarterly] Status Report No. 13 for Period Ending January 31, 1962," NASA Historical Reference Collection.

Assembling the Little Joe capsules. The capsules were manufactured "in-house" by Langley Research Center technicians. Three capsules are shown here in various stages of assembly. The escape tower and rocket motors shown on the completed capsule would be removed before shipping to Wallops Island. These test articles met the weight and center-of-gravity requirements of Mercury and withstood the same aerodynamic loads during the exit trajectory. NASA L59-4944.

Two seconds after the 10,000 foot barostat closes, power is supplied to the antenna fairing ejector -- located above the cylindrical neck section -- to deploy the main landing parachute and an underwater charge, which is dropped to provide an audible sound landing point indication. The ultra-high frequency SARAH radio then begins transmitting. A can of sea-marker dye is deployed with the reserve chute and remains attached to the spacecraft by a lanyard.

On landing, an impact switch jettisons the landing parachute and initiates the remaining location and recovery aids. This includes release of sea-marker dye with the reserve parachute if it has not previously been deployed, triggering a high-intensity flashing light, extension of a 16-foot whip antenna and the initiation of the operation of a high-frequency radio beacon.

If the spacecraft should spring a leak or if the life support system should become fouled after landing, the astronaut can escape through this upper neck section or through the side hatch.[13]

This sequence of operations worked effectively on all six of the Project Mercury piloted missions as well as during all tests.

Recovery from Long-Duration Earth-Orbital Flight: Project Gemini

In the fall of 1961, NASA began Project Gemini as a means of bridging the significant gap in the capability for human space flight between what Project Mercury achieved and what would be required for a lunar landing—already an approved program. NASA closed most of the gap by experimenting and training on the ground, but some issues required experience in space. This requirement became immediately apparent in several major areas, including the following major mission requirements, as defined in the Gemini crew familiarization manual:

- Accomplish 14-day earth orbital flights, thus validating that humans could survive a journey to the Moon and back to Earth.
- Demonstrate rendezvous and docking in Earth orbit.
- Provide for controlled land landing as the primary recovery mode.

---

13. NASA News Release, NO. 62-113, "MA-7 Press Kit," May 13, 1962, NASA Historical Reference Collection, available online at *http://www.scribd.com/doc/33284689/MA-7-Press-Kit*, accessed February 26, 2012.

- Develop simplified countdown techniques to aid rendezvous missions (lessens criticality of launch window).
- Determine man's capabilities in space during extended missions.[14]

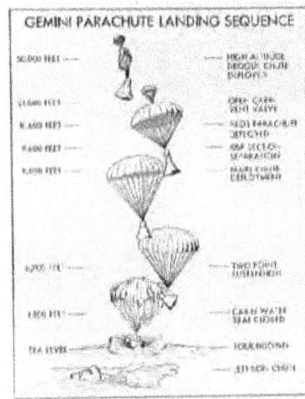

GEMINI PARACHUTE LANDING SEQUENCE

These major initiatives defined the Gemini program and its 10 piloted space flight missions conducted from 1965 to 1966.[15]

NASA originally conceived of Project Gemini as a larger Mercury Mark II capsule, but soon it became a totally different vehicle that could accommodate two astronauts for extended flights of more than 2 weeks. It pioneered the use of fuel cells instead of batteries to power the ship and incorporated a series of hardware modifications. Its designers also toyed with the possibility of using a paraglider that was being developed at Langley Research

Artist concept of Gemini parachute landing sequence from high-altitude drogue chute deployment to jettisoning of parachute. NASA S65-05398.

Center for ground landings instead of splashdowns in water and recovery by the Navy.[16] The whole system was to be powered by the newly developed Titan II launch vehicle, another ballistic missile developed for the Air Force. A central reason for this program was to perfect techniques for rendezvous and docking, so NASA appropriated from the military some Agena rocket upper stages and fitted them with docking adapters to serve as the targets for rendezvous operations.

As it turned out, the parachute system that NASA used for Gemini was similar to that used on Mercury, but it had some additional features. As one analysis noted during the program:

---

14. NASA Flight Crew Operations Division, "Gemini Familiarization Package" (August 3, 1962), NASA Historical Reference Collection.

15. The standard work on Project Gemini is Barton C. Hacker and James M. Grimwood, *On the Shoulders of Titans: A History of Project Gemini* (Washington, DC: NASA SP-4203, 1977). See also David M. Harland, *How NASA Learned To Fly in Space: An Exciting Account of the Gemini Missions* (Burlington, Ontario: Apogee Books, 2004).

16. Barton C. Hacker, "The Idea of Rendezvous: From Space Station to Orbital Operations, in Space-Travel Thought, 1895–1951," *Technology and Culture* 15 (July 1974): 373–388; Hacker and Grimwood, *On the Shoulders of Titans*, p. 126.

A single-parachute landing system is used on Gemini spacecraft, with the ejection seats serving as a backup. In the normal landing sequence, an 8-foot-diameter drogue parachute is deployed manually at approximately 50,000 feet altitude. Below 50,000 feet, this drogue provides a backup to the reentry control system for spacecraft stabilization. At 10,600 feet altitude, the crew initiates the main parachute deployment sequence, which immediately releases the drogue, allowing it to extract the 18-foot-diameter pilot parachute. At 2.5 seconds after sequence initiation, pyrotechnics release the recovery section, to which the pilot parachute is attached and in which the main parachute is stowed. As the reentry vehicle falls away, the main parachute, an 84-foot-diameter ring-sail, deploys. The pilot parachute diameter is sized such that recontact between the recovery section and the main parachute will not occur during descent. After the crew observes that the main parachute has deployed and that the rate of descent is nominal, repositioning of the spacecraft is initiated. The spacecraft is rotated from a vertical position to a 35° nose-up position for landing. This landing attitude reduces the acceleration forces at touchdown on the water to values well below the maximum which could be tolerated by the crew or by the spacecraft.[17]

As NASA engineers once noted, "Main-parachute deployments take place in full view of the crew, and it is quite a beautiful and reassuring sight."[18] At landing, the crew experienced a shock, as the "amount of wind drift, the size of the waves, and the part of the wave contacted also vary the load. Even the hardest of the landings has not affected crew performance."[19]

Interestingly, Project Gemini leaders had an entirely different approach to landing planned for this second piloted space flight program for the United States. Dangling from beneath a parachute and being rescued at sea might have been acceptable for the Mercury program, since winning the space race was an overriding motivation. However, NASA found distasteful the prospect of continuing this parachute-recovery model indefinitely. For that reason, the Gemini project managers conceived and for several years relentlessly pursued an inflatable paraglider modeled on the airfoil invented by NASA engineer

---

17. *Gemini Midprogram Conference, Including Experiment Results* (Washington, DC: NASA SP-121, 1966), p. 18.

18. Ibid., p. 276.

19. Ibid.

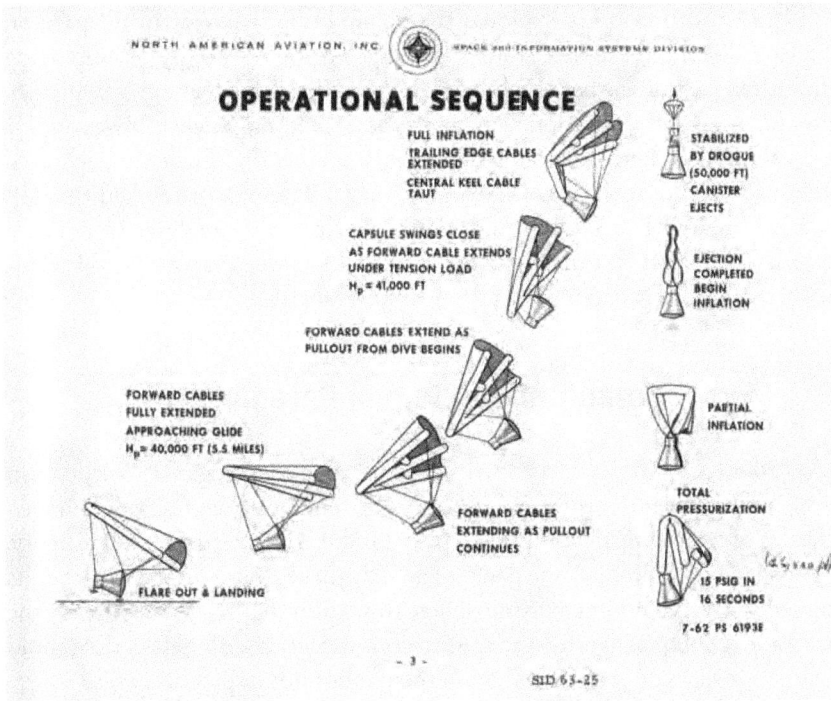

This is the proposed flight profile for the Gemini spacecraft paraglider concept in 1962. NASA.

Francis M. Rogallo, which eventually spawned the hang-gliding movement of the latter half of the 20th century. NASA's objective was a controlled, guided, soft landing on a land target of its choice. NASA human space flight head Brainerd Holmes stated in 1962:

> I think that the significant thing is that we would like to land on land, and not on a hostile sea. I think it is significant that we are planning, by having a L/D something greater than zero—as Dr. [Joseph] Shea said, .5—...to be able to guide this vehicle, if you will, just by controlling his attitude and thus having this offset center of gravity, to a localized landing area which might be an area ten miles on a side, something like that, and then much more localized through a parachute or paraglider.[20]

---

20. NASA Manned Spacecraft Center Fact Sheet, "Manned Space Flight Comes of Age as Project Mercury Nears Its End" (January 1962), p. 23, NASA Historical Reference Collection.

NASA had previously explored the possibility of recovery from space on the Rogallo wing. The first studies investigated the possibility of employing an inflated paraglider that would reenter the atmosphere at about 5,000 mph, after inflation and separation from an Aerobee sounding rocket. It was on this Inflatable Micrometeoroid Paraglider (IMP) project that flight tests achieved enough success to warrant continued research and development. Additionally, in 1963 the Air Force supported work on a reentry paraglider to rescue a crewmember from an orbiting space station. This program was called FIRST, or Fabrication of Inflatable Reentry Structures for Test.[21]

Pursuit of the Gemini Paraglider

For Project Gemini, efforts to develop a paraglider began in earnest on November 20, 1961. Through 1964, NASA engineers aggressively pursued the idea of such a practical landing system for the Gemini spacecraft. In theory, the spacecraft would carry the paraglider safely tucked away through most of its mission. After reentering the atmosphere from orbit the crew would deploy the wing and, having converted the spacecraft into a makeshift glider, they could fly to a runway landing. It worked in theory, but not in practice.

The basics of the story are well known.[22] At the start of the Gemini program in 1961, NASA considered having the two-person Gemini capsule land on a runway after its return from space, rather than parachute into the ocean. This controlled descent and landing was to be accomplished by deploying an inflatable paraglider wing. First, NASA built and tested the Parasev, a single-seat, rigid-strut parasail, designed much like a huge hang glider, to test the possibility of a runway landing. The space agency then contracted with North American Aviation to undertake a design, development, and test program for a scaled-up spacecraft version of the concept. A full-scale, two-pilot Test Tow Vehicle (TTV) was also built to test the concept and train Gemini astronauts for flight. The TTV tested maneuvering, control, and landing techniques. A helicopter released the TTV, with its wings deployed, over the dry lakebed at Edwards AFB, CA, where it landed safely. Scale models of the capsules released at higher altitudes and faster speeds sought to duplicate reentry conditions. The system never worked well enough to use on the Gemini program (largely

---

21. "Space: Rescue in Orbit," *Time* (February 1, 1963): 56.

22. The basics of this story are told in Barton C. Hacker, "The Gemini Paraglider: A Failure of Scheduled Innovation, 1961–64," *Social Studies of Science* 22 (Spring 1992): 387–406.

On June 26, 1959, then-Langley researcher Francis Rogallo examined the Rogallo wing in the Langley 7-by-10-foot wind tunnel. Originally conceived as a means of bringing piloted spacecraft to controlled, soft landings, Rogallo's concept was avidly embraced by later generations of hang-gliding enthusiasts. NASA L59-4345.

because of control and stability issues), but the Paraglider Landing System Program proved useful in developing alternate landing techniques.[23]

But there is more to the story. In the late 1940s and early 1950s, Francis Rogallo, an original thinker and kite-flying enthusiast working in the Langley Memorial Aeronautical Laboratory's 7-by-10-foot Tunnel Branch, experimented with the idea of a flexible wing that could enable individuals to fly. Working with his wife, Gertrude, they developed the familiar V-shaped flexible wing that is ubiquitous in hang gliding today, receiving patents for their

---

23. C.E. Libby, "Deployment of Parawings for Use as Recovery Systems," *Journal of Spacecraft and Rockets* 2 (1965): 274–275; W.H. Eilertson, "Gliding Parachutes for Land Recovery of Space Vehicles Case 730" (September 8, 1969), NASA-CR-108990, NASA Center for Aerospace Information, Hanover, MD.

work in 1948 and 1952. They initially focused on paragliders that resembled sailboat designs, but they quickly moved to concepts that had more in common with parachutes. When the Rogallos read the *Collier's* series of articles on space flight beginning in 1952, they immediately recognized a use for their paraglider as a method for landing when returning from space. At a 1963 American Astronautical Society conference, Rogallo said, "I thought that the rigid-winged gliders might better be replaced by vehicles with flexible wings that could be folded into small packages during the launching."[24] He pursued this idea for more than a decade before having much success in interesting space program officials in putting money into it. For Rogallo, it made more sense to fly home rather than splash down in the ocean, and his NASA colleagues in the space research-and-development (R&D) community agreed. They had recognized that the ballistic capsule dangling from the bottom of a parachute was an acceptable solution to landing, but it was also an inelegant one few wanted to use indefinitely. The paraglider concept was tailor-made to enable a step beyond the state of the art.[25] It also satisfied the space agency's desire to "advance spacecraft technology," a mission viewed as critical, even as NASA undertook operational activities in space.[26]

NASA contracted with North American in 1961 to develop a paraglider-recovery system for Gemini. As the spacecraft fell through the atmosphere back to Earth, its ablative heat shield would protect the passengers and the machine from harm while slowing it to subsonic speeds. Then, a carefully designed and packed paraglider stowed in the spacecraft would be deployed beginning at about 60,000 feet, and by 20,000 feet the descending spacecraft would take on the characteristics of a hang glider, and the astronauts would bring the craft to a controlled landing on either water or land. "In this application the wing is stowed within the spacecraft until after the high-temperature reentry and descent to low altitude, about 50,000 feet, where the speed is subsonic," states a 1963 study of the project. "At this point the wing is deployed and the pilot directs the vehicle toward a predetermined landing spot by means of manual control in pitch and roll. The pilot executes a flare maneuver at an altitude of some 100 feet above

24. F.M. Rogallo, "Parawings for Astronautics," paper presented to the American Astronautical Society Conference on Space Rendezvous, Rescue, and Recovery, Edwards AFB, CA, September 10–12, 1963, p. 1.

25. James M. Hansen, *Spaceflight Revolution: NASA Langley Research Center from Sputnik to Apollo* (Washington, DC: NASA SP-4308, 1995), pp. 380–386.

26. Kenneth L. Suit, John W. Kiker, and James K. Hinson, "Landing Rocket—Gliding Parachute Systems for Manned Spacecraft," paper presented at the AIAA Entry Technology Conference, Williamsburg-Hampton, VA, October 12–14, 1964, p. 2, Johnson Space Center Records, Record Group 255.

The Gemini paraglider was intended to enable NASA astronauts to fly back and land on solid ground rather than be rescued at sea by the U.S. Navy. NASA.

the ground, and the spacecraft lands at a low sink rate." Skids from the spacecraft would serve as landing legs for the crew returning from space.[27]

Should this system not work as intended, the Gemini spacecraft would have ejection seats and crew parachutes—which also would have helped crews escape during launch if something went wrong—that would ensure the safe return of the crew. "The use of this system may also be necessary when aborted missions result in recovery at some point where landing conditions are unsuited for use of the parawing landing system."[28] As conceptualized in 1962 at the start of

27. C.R. Foulders and G.M. Minott, "The Application of the Paraglider to Spacecraft Recovery," p. 1, National Aeronautic and Space Engineering and Manufacturing Conference, September 23–27, 1963, Johnson Space Center Records, Record Group 255, Ft. Worth Federal Records Center, TX. See also Raymond R. Clarence, Aeronautical Research Engineer, Flight Dynamics Branch, NASA Space Task Group, Memorandum for Associate Director, "Rogallo Kite," January 3, 1961, Johnson Space Center Records, Record Group 255.

28. William D. Armstrong, Flight Activities Section, Manned Spacecraft Center, Memorandum for Chief, Flight Operations Division, "Review of the Development Effort of the Parawing Landing System for the Gemini Mission" (February 9, 1962), Johnson Space Center Records, Record Group 255.

the R&D program, the optimum reentry and landing phase took place in the following way:

1. For normal recovery the parawing would be employed for the final phase of letdown and landing. The parawing is initially released at approximately 35,000 feet and is fully deployed at 20,000 feet. A gliding descent is maintained below this altitude which can be pilot-controlled to achieve landing at a preselected touchdown site.
2. Control is provided in both pitch and roll by utilizing the same hand control device used during reentry.
3. Glide is terminated at about 500 feet at which point the glide angle is increased to build up velocity for the flare maneuver. Flare is initiated at approximately 70 feet altitude, and the spacecraft is brought into the landing attitude with a rate of descent less than 5 ft/sec and a forward velocity of about 50 ft/sec.
4. The high-drag rear-landing skids and a low-drag forward skid are employed for touchdown to provide stability during the landing slide.
5. Instrumentation for control of the parawing would make use of the regular rate-and-attitude instruments used during other phases of flight. Additional instrumentation would include an airspeed meter, rate-of-descent indicator, and an altimeter.
6. The ejection seat/crew parachute backup system serves as an emergency escape system in the event of a malfunction in the parawing landing system. The use of this system may also be necessary when aborted missions result in recovery at some point where landing conditions are unsuited for use of the parawing landing device.[29]

NASA engineers pursued this option aggressively, and it is important to acknowledge how exhaustively they proceeded with the paraglider R&D program. The program engineers died very hard regarding this program once it proved untenable; in the various engineering memoranda, they talked openly about the many problems the program encountered but also about the possibilities of success. Some of those problems encountered in test operations include the following:

1. Leakage of the inflatable structure.
2. Failure of the structure to withstand the desired internal pressure and consequently blow out.
3. Malfunction of the pressurization attack release.
4. Malfunction of the attack release hook.

---

29. Ibid.

5. Rupture of the apex causing the pressurization valve to be torn from the wing.
6. Breakage of shroud lines due to snatch loads.
7. Spiral descent to impact as a result of entangled shroud lines.
8. Awkward deployment causing rotation of the spacecraft and wing.
9. Buckling of the wing due to the ratio of dynamic pressure to internal pressure.
10. Nonrelease of the drogue chute during deployment.
11. Mechanical failure of activating devices and electrical systems.

Other problems that had not yet been demonstrated but were certainly possible also prompted engineers to pursue solutions. "From the pilot's point of view there is a big question as to whether the complicated control system will provide the pilot with sufficient control of the vehicle during all possible flight conditions such as wind gusts, precipitations, and limited visibility," the engineers acknowledged. They were especially concerned about the performance of the spacecraft and the paraglider during return to Earth in harsh weather conditions. "If high winds are encountered," they stated, "then it is very likely that the system will collapse or tumble and become entangled with the shroud lines." Engineers also wanted to ensure that the spacecraft could come down in water as well as on land, in case that was necessary during an emergency. They also sought a release mechanism from the paraglider to allow use of the ejection system should an emergency arise at any point during recovery. Finally, NASA engineers worked to ensure that the paraglider would deploy appropriately under all conditions.[30]

Because of these problems with the paraglider system, the Gemini project manager initiated a "plan B" for recovery: the development of a parachute system similar to that used in Project Mercury.[31] Reasonable confidence abounded that this system, proven effective, could also work for the Gemini program. Gemini design engineers also toyed with a parasail concept—a canopy that looked more like a parachute than a paraglider but had modest maneuverability. NASA contracted with Pioneer Parachute Company for the testing of the parasail concept in 1963. The parasail concept was something of a cross between the parachute and the paraglider. It had the following char-

30. James M. Rutland, Aerospace Technologist, Manned Spacecraft Center, Memorandum for Chief, Flight Operations Division, "Recovery System for Gemini," March 21, 1962; G.M. Moisson, North American Aviation, to S. Kriedel, "Trip Report—Materials Survey, Paraglider Fabrics and Castings," October 9, 1962, both in Johnson Space Center Records, Record Group 255.
31. James A. Chamberlin, Manager, Project Gemini, "Project Gemini Abstract of Meeting on Backup Parachute Program," July 14, 1962, Johnson Space Center Records, Record Group 255.

This test used a Volkswagen Beetle as the platform to test aerodynamics of the Gemini paraglider concept. NASA.

acteristics, as noted in a 1962 proposal for the R&D program from Pioneer Parachute Company.

1. It is constructed of very low porosity fabric.
2. It is designed to wing and airfoil theories.
3. It uses slots and vents to relieve excess pressures and to guide air into and out of the canopy.
4. It makes use of central suspension lines to achieve high drag.
5. Originally this canopy was *not* designed for deployment at any appreciable speed.
6. It has been built in one size (23.2 ft equiv. nom. Dia.) only.
7. Little is known of opening forces and canopy stresses.[32]

With challenges before it, NASA and contractor engineers believed that this approach had potential for future space vehicle recovery and landing.

As a 1965 report noted: "The total Parasail land-landing system includes the gliding parachute for local obstacle avoidance and landing attenuation rockets fired just above the surface to reduce descent velocity prior to impact.

---

32. Pioneer Parachute Co., "Technical Proposal for Para-Sail Evaluation and Development," NASA RFP Number MSC-83-239P (November 23, 1962); Leland C. Norman and Jerry C. Cofrey, "Development Tests of the 18 Ft Diameter Para-Sail," Proposed NASA MSC Technical Note, n.d., both in Johnson Space Center Records, Record Group 255.

The program we have been conducting has included component development of the parachute, the landing attenuation rockets, turn control motors for steering, altitude sensors, pilot display and visual reference system, and landing gear." NASA successfully flew this system four times, and while it was developed too late for the Gemini program, many believed it could be incorporated into the Apollo recovery system.[33] Starting April 21, 1965, a test program capsule, dubbed "El Kabong," was dropped from an Air Force Reserve C-119 from an altitude of 11,500 feet at Fort Hood, TX. During descent tests, personnel steered the parasail by radio command to operate motors on the capsule that controlled flap angles on the sail and allowed modest control of drift. As the test spacecraft neared the ground, sensors ignited two 6,000-pound thrust motors that reduced capsule speed to less than 10 ft/sec. The capsule then landed on tricycle landing gear. The first two tests did not go as planned, but on the third, flown on July 31, 1965, the capsule landed within 40 feet of its target at Fort Hood's Antelope Mound tank range. "We've got a winner!" an engineer named Lee Norman announced at the conclusion of the test. He announced correctly (as all earlier tests had taken place over water), "This is the first successful landing [of a spacecraft] in this country!"[34]

However, the parasail was not the recovery system of choice, rather it was plan B; NASA's goal was to successfully integrate the inflatable paraglider. To prove the concept, the space agency first developed and tested the Parasev, the first one of which was tested in Langley's full-scale wind tunnel in January 1962. Parasev-1 used a hang-glider-type steel-tube frame onto which the fabric was fixed. The powerless, lightweight steel-tube vehicle was taken to what would become the Dryden Flight Research Center at Edwards AFB for flight tests. It was towed by either a ground vehicle or a light airplane to altitudes of up to 12,000 feet for a free-glide, and then it came in for a 100-mph landing on the dry lakebed. Early flights with experienced test pilots (Gus Grissom, Milton O. Thompson, and Neil Armstrong among them) proved extremely

---

33. George E. Mueller, NASA Associate Administrator for Manned Space Flight, to NASA Associate Administrator, "Parasail and Paraglider Effort" (August 13, 1965); "Parasail—Retrograde Landing System Study" (May 1, 1963); "Structural Design Criteria (Parasail)" (June 24, 1963); Charles W. Mathews, Manned Spacecraft Center, "Project Gemini Abstract of Meeting on Paraglider Landing System" (September 11, 1963), all in Johnson Space Center Records, Record Group 255.

34. Clay Coppedge, "Fort Hood, 'El Kabong' Footnote in Space Exploration," *Temple Daily Telegram* (Temple, TX), July 23, 2007; Leland C. Norman, Robert B. West, and David L. Brown, "Development of the Para-sail Parachute as a Landing System for Manned Spacecraft," paper presented at Symposium on Parachute Technology and Evaluation, El Centro, CA, April 7–9, 1964, Johnson Space Center Records, Record Group 255.

tricky. One pilot noted it flew as if "controlled by a wet noodle." During one ground tow, the pilot got out of phase with a lagging control system and the craft developed a rocking motion that grew worse the more he tried to correct for it. It was a version of pilot-induced oscillations, which the Space Shuttle would experience a decade and a half later. Parasev-1 did a wing over into the lakebed, demolishing the craft and injuring the pilot. It was rebuilt several times and eventually made over 100 flights.[35]

Eight of NASA's Flight Research Center's engineers conducted this test program under the direction of Charles Richards. Its name, Parasev, an abbreviation of Paraglider Research Vehicle, served as useful shorthand for a program that the public saw as cute but also useful. Parasev-1 resembled an unpowered tricycle with an angled tripod mast, on top of which sat the Rogallo wing. Constructed totally in-house, Parasev-1 received Federal Aviation Agency (now Federal Aviation Administration) approval on February 12, 1962. The vehicle itself was unremarkable; it was an open frame "fuselage" made from welded 4130 titanium tubing, with aluminum tubing comprising the keel and wing leading edges. The pilot sat in the open, strapped to a seat with no enclosure of any kind. The pilot could control the vehicle by tilting the Parasev wing fore and aft, and turn it by tilting the wing with a control stick. The control stick, in contrast to a normal aircraft's, was mounted overhead in front of the pilot's seat. The wing, made of Dacron and linen, was sown for Parasev-1 by a sailmaker.[36]

The Parasev proved a difficult flying vehicle to master. Tow tests behind a utility truck first got the vehicle airborne. Later, "souped-up" cars, motorcycles, and finally a helicopter towed the Parasev-1. One writer described the process of flying the vehicle as follows:

> Floating it along the ground while getting to know the handling characteristics before attempting more ambitious manoeuvres, it was not an untroubled learning curve. Flapping violently in the

---

35. Robert R. Gilruth, NASA Space Task Group, to George M. Low, "Development program on spacecraft landing systems," July 6, 1961, NASA Langley Research Center Historical Files, Hampton, VA; Richard P. Hallion, *On the Frontier: Flight Research at Dryden, 1946–1981* (Washington, DC: NASA SP-4303, 1984), pp. 137–140; James R. Hansen, *Spaceflight Revolution*, pp. 385–387; Michael Collins, *Liftoff: The Story of America's Adventure in Space* (New York: Grove Press, 1988), pp. 82–83.

36. Floyd L. Thompson to NASA, Code RAA, "Proposed Research Authorization entitled 'Free-Flight and Wind-Tunnel Tests of Guided Parachutes as Recovery Devices for the Apollo Type Reentry Vehicle'," August 30, 1961, NASA Langley Research Center Historical Files; "The Rogallo Parasev: A Revolution in Flying Wings," *Aviation News*, March 2007, available online at *http://www.aviation-news.co.uk/Parasev.html*, accessed September 11, 2009.

The Parasev-1-A (Paraglider Research Vehicle) and the tow airplane (450-horse power Stearman sport biplane) sitting on Rogers Dry Lake, CA. The control system in the Parasev-1-A had a conventional control stick position and was cable-operated; the main landing gear used shocks and bungees with the 150-square-foot wing membrane made of 6-ounce unsealed Dacron. NASA E-8712.

wind, the lobes bulging and straining in a most alarming way, the linen wing was giving problems, including flutter at the trailing edges. Longitudinal and lateral stick forces were considerable. Flying the Parasev was, said [NASA research pilot Bruce] Peterson, more difficult than the later, more sophisticated, Lifting Body series of unpowered gliding vehicles. Several changes were made to the rigging arrangement and control modifications were tested but few responses were effective and none were predictable. On the fifth flight aloft, Peterson got out of phase with the control lag and a sinusoidal wallowing motion set in, but at the moment the tow-truck began to slow down the Parasev flipped over and crashed to the desert floor from a height of 10ft, just 10sec after lift-off. Peterson survived intact but the Parasev did not, a rebuild necessitating a new designation – and a new, Dacron-covered Parawing![37]

Even so, NASA believed it had a promising concept and persisted in its development. But NASA determined by May 1963 that the Parasev technology would not be pursued any further for Gemini. Instead, Gemini leaders

---

37. "The Rogallo Parasev," available online at *http://www.aviation-news.co.uk/Parasev.html*, accessed September 11, 2009.

thought the Parasev "a basic research vehicle and not directly applicable to Gemini paraglider landing system program."[38]

As one of several design and test-development articles used to evaluate the paraglider concept, NASA built a half-scale test vehicle (HSTV) for dropping out of a helicopter and testing the paraglider. The first test took place at NASA's Flight Research Center in the Mojave Desert on December 10, 1962. It was an inauspicious performance. As the test report stated:

> The test vehicle was launched at approximately 1400 PST from 12,300 foot pressure altitude (indicated) and a velocity of 51 knots indicated airspeed which was in accordance with the launch conditions specified in the Detailed Flight Plan.... Immediately following release, the test vehicle tumbled into the riser lines of the drogue parachute. This tumbling apparently resulted from the break-ties used for stabilization during pre-launch flight.
> Shortly after the drogue chute was untangled the HSTV went into a spinning condition (describing a cone).... The capsule continued to spin throughout Paraglider pressurization, aft restraint release and apex release. This spinning condition prompted the decision to jettison the Paraglider and deploy the parachute.[39]

The HSTV landed safely about 120 seconds after deployment, but the test demonstrated one of the significant problems of the paraglider throughout its R&D experience. The next test, on January 7, 1963, had to be aborted because of radio-command system failures. They tried again the next day, this time with disastrous results. "At an estimated 1000 feet from the ground, and the absence of the main chute, the emergency chute signal was transmitted from the Comm. Van with no effect," the test report states. "The capsule was damaged severely and beyond repair."[40] Tests followed on May 22 and June 3,

---

38. Charles M. Mathews, Manned Spacecraft Center, "Project Gemini Abstract of Meeting on Paraglider Landing System" (May 21, 1963), Johnson Space Center Records, Record Group 255.

39. M.S. Jackson, North American Aviation, "Preliminary Flight Test Report on Flight 6-3A-B/b (First HSTV Deployment)" (December 13, 1962), Johnson Space Center Records, Record Group 255.

40. M.S. Jackson, North American Aviation, "Report on Flight Operations on Monday 7 January 1963" (January 7, 1963); G.J. Zavadil, North American Aviation, "Preliminary On-Board Paraglider Data for Test Number 7-3A-A/d" (January 14, 1963), with attachments; and R.B. Bartle, "Summary Report on Failure Analysis, HSTV Flight Operation No. 7-3A-A/d" (February 12, 1963), all in Johnson Space Center Records, Record Group 255.

1963, but during a July 30 drop, both the main and backup parachutes failed and the vehicle was destroyed.[41]

In the end, the HSTV flights demonstrated a pattern that was replicated throughout the entire effort. The unfurling of the paraglider never seemed to proceed as intended and, sometimes, emergency systems failed to compensate. There was certainly no way NASA could put astronauts into a vehicle with such complications. NASA's Gemini Project Manager James A. Chamberlin repeatedly called into question the entire effort, but he was dissuaded from cancelling it, at least for the time being. He told North American senior official Harrison A. "Stormy" Storms, Jr., on September 21, 1962, that there was "growing concern" over "repeated unsuccessful attempts...to conduct satisfactory predeployed half-scale paraglider tests." Because of this, Chamberlin ordered a standdown for the test flights until North American could restructure its test program and give reasonable assurance that it could proceed effectively.[42]

Considerable changes to the system followed, and in October 1962 the final half-scale test flight achieved its main goal of demonstrating stability in the paraglider system during free flight.[43] The next three flights suggested that the October 1962 test might have been a fluke. The first two of these were only partially successful; the third, on March 11, 1963, failed to deploy the wing, and then the emergency backup parachute failed as well. With this crash, two HSTV test articles were too much wreckage. The testing ended and Chamberlin, who had been skeptical about this for some time and wanted to end the paraglider program, lost his job and was replaced by a new Gemini program manager committed to the new landing concept.[44] NASA reorganized the HSTV test effort once again, proposing a two-part test program in which a capsule would be deployed from an aircraft and then deployment of the wing would be tested in flight. Upon completing that successfully, North American and NASA personnel would tackle the second phase: landing the vehicle safely.

---

41. Hacker and Grimwood, *On the Shoulders of Titans*, pp. 144–146.
42. James A. Chamberlin, NASA, to Harrison A. Storms, North American Aviation, "One-Half Scale Paraglider Program" (September 21, 1962), NASA Historical Reference Collection.
43. Charles W. Mathews, Manned Spacecraft Center, "Project Gemini Abstract of Meeting on Parachute Landing System" (October 9, 1963), Johnson Space Center Records, Record Group 255.
44. James A. Chamberlin to Gemini Procurement Office, "Phase III, Part 2 of the Gemini Paraglider Landing System Program" (February 20, 1963); Charles W. Mathews, Manned Spacecraft Center, "Project Gemini Abstract of Meeting on Paraglider Landing System, March 13, 1963" (March 15, 1963), and A.A. Tischler to James A. Chamberlin, "Preliminary Flight Test Report—Half-Scale Paraglider Deployment Test No. 3, Contract NAS 9-167" (March 14, 1963), all in Johnson Space Center Records, Record Group 255.

NASA engineers recovering the Gemini paraglider after a test in the Mojave Desert. NASA.

That second task involved the use of two full-scale trainers towed to altitude and piloted down to the surface. Once this was mastered, according to the research proposal, NASA could move on to test the entire sequence.[45]

NASA never got to that stage. The half-scale tests ended much later than anticipated, and demonstrations of the ability to deploy the parawing in flight were not completed. It seemed there was virtually no way to get this system ready to fly for Gemini, even 2 years before the first orbital flights of the spacecraft. Some still wanted to cancel the effort, but North American pressed for its continuation, saying that the problems had been resolved, and the paraglider test program was granted a stay of execution.[46]

Throughout this tough cycle of R&D, few other than Chamberlin truly wanted to give up on the paraglider concept, and even Chamberlin would have preferred to make it work. Several people weighed in to preserve the R&D effort. For example, Major General Ben I. Funk, commander of the U.S. Air Force Space Systems Division, saw a need for the landing system and was willing to support it, despite it not being a military program. "I would not like to see the paraglider effort dropped," he wrote to Manned Spacecraft Center Director Robert R. Gilruth on September 16, 1963. "[F]or all the procedural and technical problems which had beset its development so far, the paraglider may still prove the best way of land-landing a low lift/drag spacecraft, a capability in which we are interested for possible future military operations."[47]

To prove the second phase of the full-scale parawing recovery and landing system, NASA built an "advanced paraglider trainer" for use by astronauts preparing for Gemini flights. It was modified in 1963 to serve as the first of two TTVs. The first, TTV-1, was used in perfecting maneuvering, control, and landing techniques. It had wheeled landing gear, but NASA also considered the use of landing skids. In essence, this vehicle was a boilerplate of the Gemini capsule that had the same dimensions as the flight article. "The cockpit floor, side bulkheads and substructure below the floor were simulated to permit installation of Gemini landing gear at a later point in time. The frames and

---

45. Hacker, "The Gemini Paraglider System," p. 394.

46. Charles W. Mathews, Manned Spacecraft Center, "Abstract of Meeting on Paraglider Landing System, February 6, 1963" (February 8, 1963); Charles W. Mathews, Manned Spacecraft Center, "Abstract of Meeting on Paraglider Landing System, February 13, 1963" (February 18, 1963); and James A. Chamberlin to Robert L. Kline, "Continuation of Phase II, Part A, Paraglider Development Program, Contract NAS 9-167" (February 20, 1963), all in Johnson Space Center Records, Record Group 255.

47. Maj. Gen. Ben Funk, USAF Space Systems Division Commander, to Robert R. Gilruth, Manned Spacecraft Center Director, September 16, 1963, Johnson Space Center Records, Record Group 255.

This flight test of a 50-foot parawing's ability to bring down safely a model of a human space capsule from a few thousand feet took place on an old army bombing range near Langley Field, VA, in 1961. NASA L61-8041.

bulkheads were of welded tri-ten steel with nonstructural and structural doors of aluminum plate for access to internal equipment."[48] The vehicle always had difficulties with control, especially the "flare" that allowed for a soft landing. For flight tests, TTV-1, and later TTV-2, flew with its paraglider wing already deployed; it was towed aloft and released. TTV-1 performed eight of these drop tests and landings at the Edwards dry lakebed.[49]

The first of these piloted TTV tests, albeit with the paraglider already unfurled, took place on July 29, 1964, when a helicopter towed TTV-1 with

---

48. N.F. Witte, North American Paraglider Program Manager, "Final Report of Paraglider Research and Development Program, Contract NAS 9-1484, SID65-196" (February 19, 1965), North American Aviation, Johnson Space Center Records, Record Group 255.

49. "Gemini Paraglider Flare Problem," n.d.; Goodyear Aircraft Corp., "Design Summary Report, Feasibility Study Advanced Concept—Inflatable Paragliders," GER-10846 (September 21, 1962); and James A. Chamberlin, Manned Spacecraft Center, "Project Gemini Abstract of Meeting on Paraglider Landing System" (November 9, 1962), all in Johnson Space Center Records, Record Group 255.

North American Aviation test pilot Charles "E.P." Hetzel aboard. He released from the helicopter at approximately 10,000 feet and then undertook a descent to the ground that took more than 20 minutes. A second free flight, on August 7, went well until at about 3,000 feet, when TTV-1 went into a hard left turn and headed for the ground; Hetzel bailed out and suffered modest injury on landing. TTV-1 was damaged during the test, and there were no further piloted flights of the vehicle until after 14 remotely controlled tests had been completed. A flight on December 19, 1964, also ended badly, with research pilot Donald F. McClusker flaring the paraglider late and suffering injuries in the hard landing. Persistence paid off, however, and eventually NASA was able to complete free flights on the TTV-1—but not before the program was divorced from the Gemini program.[50]

In addition to the piloted TTV effort, a nonpiloted, full-scale test vehicle (FSTV) underwent drop tests from the back of a C-130 aircraft. As the final report noted, this "vehicle was designed to withstand loads and stresses peculiar to the test trajectories with sufficient conservatism to preclude catastrophic failures during flight."[51] North American built two FSTVs and designed a set of drop tests to see how best to ensure the paraglider deployment sequence. At about 10,000 feet the test vehicle would then jettison the paraglider in favor of landing on the emergency parachute system.[52] Twenty tests were scheduled, but the program conflicted with gravity almost from the beginning. The first test, on January 22, 1964, failed when the paraglider failed to deploy. Three more tests, February 8, March 6, and April 10, also failed to yield positive results. Then, after yet another failure of an April 22 test, NASA managers had no choice but to cancel the program. It took months to come to this decision, and Gemini program officials did not publicly announce the decision until August 10, 1964, allowing North American to continue this effort as an R&D program unrelated to Project Gemini until the contract expired at the end of the year.[53]

The R&D program for the paraglider extended from May 1962 until December 1964. The latter date was just before the inauguration of piloted flights of the Gemini spacecraft, the next spring. Indeed, as the program proved less successful than originally envisioned, NASA engineers kept pushing back deployment of the paraglider, suggesting that the first few missions could use

---

50. Dwayne A. Day, "A Wing and a Prayer: The Troubled Development of the Gemini Paraglider, Part 1," *Spaceflight* 49 (August 2007): 302–309; Dwayne A. Day, "Failures and Setbacks: The Troubled Development of the Gemini Paraglider, Part 1," *Spaceflight 49* (September 2007): 342–345.

51. Witte, "Final Report of Paraglider Research and Development Program."

52. Hacker and Grimwood, *On the Shoulders of Titans*, p. 133.

53. Hacker and Grimwood, *Project Gemini: A Chronology*, pp. 92–93, 120.

This technical drawing
shows the configuration of
the paraglider for stowage
and flight tests. NASA.

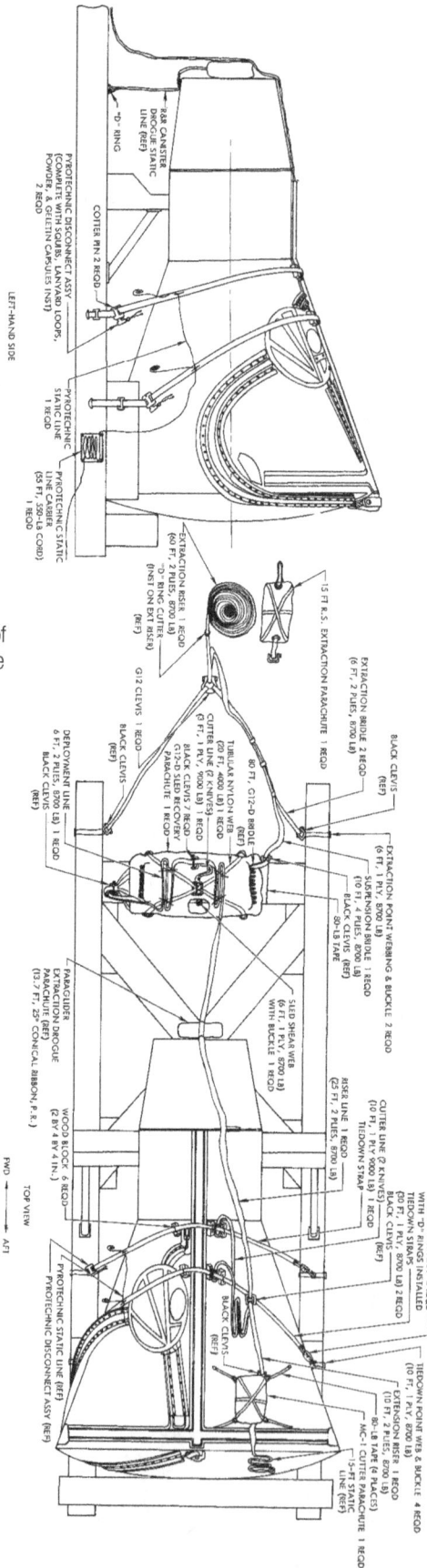

Figure 15. FSTV Vehicle/Sled Tie-Down System

SID 65-196

conventional parachutes but later flights would incorporate the paraglider. At one point in 1964, NASA wanted to have the first seven Gemini capsules use a traditional parachute-recovery system and the last three missions employ the paraglider. This proved a pipe dream as well, and soon the program's sole remaining goal was completion of a technology development effort with no relationship to Project Gemini. A synopsis of the test program was contained in the paraglider development effort's final report:

> The program incorporated a series of half-scale and full-scale wind tunnel tests, twenty-five deployment flight tests and four free-flight maneuver and landing tests. Ancillary testing included development of the FSTV parachute recovery system through boilerplate drop tests; half-scale free flight tests; one-tenth scale and one-twentieth scale drop tests.[54]

Test program officials concluded, somewhat ebulliently, and perhaps too generously, in the final analysis: "The flight test program was concluded with complete success in the FSTV and TTV phases."[55] In an irony of the first magnitude, after yet another failure of the program on April 22, 1964 (the fifth consecutive failure), NASA cut its losses and ended the paraglider program. On the day after NASA began phasing out the paraglider, April 30, 1964, the program finally had a successful test. Playing out the string must have lessened the pressure on the test team, as thereafter it conducted 19 more tests (out of 25 total), and in the summer it worked out the deployment sequence and began racking up successes, one after another.[56]

NASA's Administrator for Manned Space Flight, George E. Mueller, tried to explain the strange career of the Gemini paraglider R&D program to Associate Administrator Robert C. Seamans in August 1965:

> As you know, the Paraglider development program was instituted as a part of our mainstream Gemini effort in FY 1962 as a promising land-landing concept. The inflatable Rogallo wing configuration had been demonstrated in some small scale applications and the performance was certainly encouraging enough to warrant investigation for manned land-landing application. A very similar type of development approach to that followed for the Parasail

---

54. Witte, "Final Report of Paraglider Research and Development Program," p. 12.

55. Hacker and Grimwood, *Project Gemini: A Chronology*, p. 120.

56. Hacker, "The Gemini Paraglider Program," pp. 395–396.

was conducted for the paraglider, including a number of component developments, scale tests, and full-scale flight tests. This total effort in the mainstream Gemini program amounted to $27.4M (FY 1962 - $0.3M; FY 1963 - $12.1M; FY 1964 - $15M) with the bulk of the funds obligated for system development and flight test ($22.2M) and the remainder used for design and component development. Because of the inherent heavier weight encountered with the Paraglider, and the problems of spacecraft stowage and Paraglider deployment, considerable emphasis was placed throughout the development on the requirement for a light-weight configuration and a feasible and operationally suitable deployment sequence. This program culminated in a limited series of full-scale flight tests in December 1964, one of which was manned.[57]

Mueller concluded that the program, despite modest success toward its end, was only promising as a technology development program removed from the Gemini program.[58] He let the program run its course, but as Seamans agreed: "the paraglider test program will phase out with the work presently under contract. This will be concluded by December of this year. Therefore, it is understood that there will be no requirement for further funding of paraglider activities."[59]

A core question that one must ask about this program is why it failed. The first and most significant factor was the technological challenge. Deploying an inflatable structure from the capsule and gliding it to a landing on the surface is a task not without difficulties. North American was never able to overcome those difficulties. If the problem was not the deployment, it was the control mechanism. If it was not either of those, it was the difficulty in piloting the vehicle. If it was none of those, it was the size and weight of the paraglider in relation to the capacity of the Gemini capsule. Historian Barton C. Hacker concluded: "Paraglider's failure clearly owed something to intransigent technology, something to limited human and capital resources."[60] As Dwayne Day

---

57. George E. Mueller, NASA Associate Administrator for Manned Space Flight, to NASA Associate Administrator, "Parasail and Paraglider Effort," (August 13, 1965), Johnson Space Center Records, Record Group 255.

58. Ibid.

59. Earl D. Hilburn, NASA Deputy Associate Administrator, to NASA Associate Administrator for Manned Space Flight, "Parasail and Paraglider Effort" (August 19, 1965), Johnson Space Center Records, Record Group 255.

60. Hacker, "The Gemini Paraglider Program," p. 398.

concluded: "The paraglider at first seemed like a promising technology that could solve the problem of bringing a spacecraft down to a specific area and placing it gently on the ground. But the technology never matured despite considerable time and effort spent by NASA to perfect it."[61] Hacker also found fewer technical reasons for the lack of success, reasoning the program also failed because of "bureaucratic friction and internal politics, rigid schedules and competing demands for resources, clashing institutional and national priorities." He documented in-fighting, bureaucratic politics, and a host of other issues that also affected the program.[62]

Another question that might be asked about the program is why did NASA project managers persist with the paraglider system as long as they did before admitting the technology was not yet ready for prime time. There are a number of possibilities that could help explain this seeming incongruity. One factor, without question, is that no one wants to admit defeat. A program had been approved, preliminary analysis suggested this was an achievable goal, and money was available to pursue the task to a successful conclusion. No engineer could have been better motivated. The paraglider represented a useful solution to an understandable problem in a program of national priority; the program was well funded, and there was a foreseeable solution. These characteristics helped attract the Gemini program staff, both NASA and contractor employed. Naysayers who thought this untested concept was too risky to the lives of the Gemini astronauts served to steel the supporters' resolve.

There is one other factor that propelled the paraglider R&D effort: the desire to fly home like an airplane and land on a runway, or at least a skid strip, rather than bobbing in the ocean while waiting for the Navy to rescue the capsule and crew. Virtually all of the NASA engineers of the era came out of aviation engineering programs and cut their teeth on aerodynamics and hypersonics. Ultimately, people pursue design solutions that reflect their backgrounds and perspectives. A winged return to Earth had dominated the thinking about space flight from the 1920s until the ballistic era of the 1950s. The paraglider allowed an incremental step back toward what everyone believed should become the norm in space recovery. Given this historical tradition of winged space vehicles, the relatively small Mercury, Gemini, and Apollo spacecraft placed atop ballistic missiles seemed like anomalies, and reentry from space on parachutes seemed like an admission of defeat. Moreover, the primary goal of the landing system was pinpoint landings. A wing, or just aerodynamic lift (to include lifting bodies), was one way to achieve this goal. Rotors and lift rockets, like the

---

61. Day, "A Wing and a Prayer," p. 309.
62. Hacker, "The Gemini Paraglider Program," p. 398.

lunar lander, were also possibilities, as were steerable gliding parachutes, Rogallo wings, and the parafoils advocated by this program. Each method worked in theory, but the weight, storage, and deployment of these options caused serious problems during flight tests. That it took so long to realize that the paraglider concept would not be ready for Gemini's use is one of the more interesting aspects of the story. On some level, program participants wanted to believe they could be successful and move the concept to the stage of piloted space landing.

The well-respected aerodynamics researcher Bob Hoey commented on this problem for this study:

> Engineers considered all options for achieving reusability and pin-point return. Gene Love from Langley used to separate the concepts into two categories: (1) Coupled Mode – which included wings, lifting bodies, or other entry configurations which could be recovered, (landed) without deploying a special "landing" system. (2) Decoupled Mode – which included parachutes, rotors, landing rockets, or other special "landing" devices which were deployed after entry, and grossly altered the vehicle configuration and controllability as the vehicle neared the ground. Design engineers respond to the requirements that are laid on them, and are not responsible for misperceptions or errors in the requirement specifications.[63]

Historians of technology use the phrase "the social construction of technology" to symbolize decisions made not solely for engineering purposes but also for design preferences, usually resulting from psychological conceptions. At some level, the paraglider program seemed to have evolved from what its creators knew about the political, economic, social, and technological issues it addressed. It was, accordingly, a product of heterogeneous engineering, which recognizes that technological issues are simultaneously organizational, economic, social, and political. Such an outcome was understandable as people, institutions, and interests came together to launch the program. The final decision met most of the priorities brought to the process but left others unaddressed.[64]

---

63. Bob Hoey comments on manuscript, March 2011. Our thanks to him for these insights.
64. John Law, "Technology and Heterogeneous Engineering: The Case of Portuguese Expansion," in *The Social Construction of Technological Systems: New Directions in the Sociology and History of Technology*, ed. Wiebe E. Bijker, Thomas P. Hughes, and Trevor J. Pinch (Cambridge, MA: MIT Press, 1987), pp. 111–134; Donald MacKenzie, "Missile Accuracy: A Case Study in the Social Processes of Technological Change," Ibid., pp. 195–222.

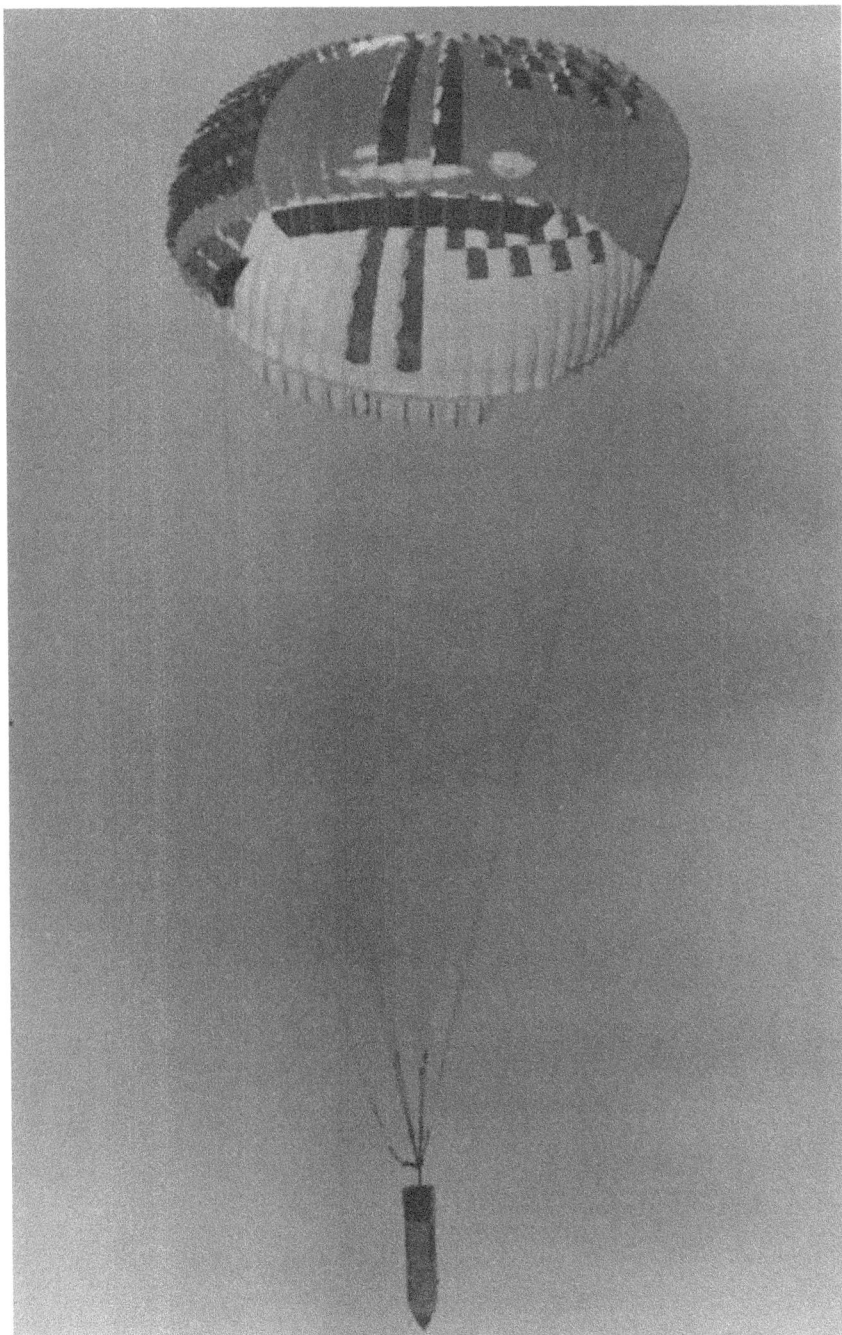

As a plan B, NASA pursued this parasail concept, which also offered a modicum of control for landing. NASA.

As one observer, Valtteri Maja, asked rhetorically, "Why not use parachutes instead of parafoils?" The answer is, "You cannot really steer them much at all. There is very little lift and mostly just drag, which means that the descent speed is quite large compared to the chute size. And it also means that you can't fly them." Maja added:

> Flying has numerous advantages compared to falling. You can do a flare move just before landing to make the landing very gentle. You might need skids or even wheels though in case the flare doesn't stop you completely. And you can navigate when you fly. Precision skydivers routinely land their feet directly on a point target in competitions. If your ram-air parafoil has a lift to drag of 10, and you start gliding at 5 kilometer height, you can theoretically pick any landing point in a 50 km radius. You can land on a runway. You can have better L/D than the space shuttle, meaning an easier and more flexible and maneuverable landing. But I'm getting carried away, this post was about the *last five kilometers* only. The main point is that the parafoil is the capsule's way out of the disadvantaged position of helpless passive falling and need for extensive recovery which has been the tradition so far. Capsules don't have to be like that. The world has moved on since the sixties and offers wonderful opportunities. With the parafoil, *capsules can fly.*[65]

Flying seems to have been the key here. What dominated the resolve with which NASA engineers pursued this concept, notwithstanding technical challenges that seemed insurmountable, ultimately was the belief that any recovery method that allowed astronauts to pilot their space vehicles to landings on Earth was superior to dangling from the bottom of parachutes and being rescued at sea by the U.S. Navy.[66]

---

65. Valtteri Maja, "The Last Five Kilometers," May 9, 2008, online blog, *Gravity Loss*, at *http://gravityloss. wordpress.com/2008/05/09/the-last-five-kilometers/*, accessed September 11, 2009.

66. Bob Hoey commented that he disagreed with this assessment. As he remarked, "The landing method was determined by solid, due-diligence engineering, not by some psychological reaction from past practices. Progress was being measured by operational considerations, and the continued goal of reusability and pinpoint recovery, and thus low operating costs for a high flight frequency. Obviously the pressure from the cold war to do it quickly, caused these long term operational goals to be deferred for Mercury, Gemini, and Apollo." The question remains unresolved. Does social construction explain the persistence of the paraglider concept or does something else?

Three Canopies into the Pacific: Project Apollo

With the failure of the paraglider program for Gemini, NASA approached recovery from space for the Apollo spacecraft using a parachute concept. Of course, the race to the Moon through Project Apollo led to sophisticated technological innovations in the relationship between humans and machines. While putting an American on the Moon in 1969 was a feat of admittedly astounding technological virtuosity, the codependent relationship of humans and the machinery of space flight, and the sophistication of mechanical systems under human control, accounted for the six successful Moon landings between 1969 and 1972. Project Apollo succeeded in no small measure because it was a triumph of management in meeting enormously difficult systems-engineering and technological-integration requirements, as several historians have pointed out; but the machines controlled by the astronauts, more than anything else, went far toward making the process work. In this process, the Apollo astronauts proved their supremacy over all of the technological systems that were created to make voyages to the Moon possible. Indeed, they became the critical component in the effort, including the landing system.

For all of the effort on the Gemini's Earth landing system—paraglider, parasail, or parachute—virtually nothing about the Apollo program revolved around its Earth landing system. Gone were the extravagant efforts to achieve a nonwater landing, as the Apollo astronauts would be rescued at sea after a parachute landing. The Apollo familiarization manual described the system used for this recovery as follows:

> The C/M-ELS begins operation upon descending to approximately 24,000 feet +0.4 second, or in the event of an abort, 0.4 second after launch escape assembly jettison…. The apex cover (forward heat shield) is jettisoned by four gas-pressure thrusters. This function is imperative, as the forward heat shield covers and protects the ELS parachutes up to this time. At 1.6 seconds later, the drogue mortar pyrotechnic cartridges are fired to deploy two drogue parachutes in a reefed condition. After 8 seconds, the reefing lines are severed by reefing line cutters and the drogue parachutes are fully opened. These stabilize the C/M in a blunt-end-forward attitude and provide deceleration. At approximately 10,000 feet, drogue parachutes are released, and the three pilot parachute mortars are fired. This action ejects the pilot parachutes which extract and deploy the three main parachutes. To preclude the possibility of parachute damage or failure due to the descent velocity, the main parachutes open to a reefed condition for 8 seconds to further decelerate the C/M. The three

parachutes are then fully opened (disreefed) to lower the C/M at a predetermined descent rate. At 27-1/2-degree hang angle of the C/M is achieved by the main parachute attachment points. In the event one main parachute fails to open, any two parachutes will safely carry out the prescribed function.

The main parachutes are disconnected following impact. The recovery aids consists of an uprighting system, swimmers umbilical, sea (dye) marker, a flashing beacon light, a VHF recovery beacon transmitter, a VHF transceiver, and an H-F transceiver. A *recovery* loop is also provided on the C/M to facilitate lifting. If the command module enters the water and stabilizes in a stable II (inverted) condition, the uprighting system is activated (manually), inflating three air bags causing the command module to assume a stable I (upright) condition. Each bag has a separate switch for controlling inflation. The sea (dye) marker and swimmer's umbilical are deployed automatically when the HF *recovery* antenna is deployed (manually initiated by crew). The marker is tethered to the C/M forward compartment deck and will last approximately 12 hours. The swimmer's umbilical provides the electrical connection for communication between the crew in the C/M and the recovery personnel in the water.[67]

The spacecraft would land in the Pacific—for all of the lunar missions—reaching the water at a velocity of "33 feet per second at 5,000 feet altitude for a normal or abort landing."[68]

To develop the Apollo landing system, NASA contracted with North American Rockwell. Building on knowledge gained in the Gemini and Mercury parachute-landing systems, North American Rockwell undertook a rigorous and extensive design and testing regimen. Northrop engineer Theodor W. Knacke reported in 1968 the following:

> Numerous interesting design details are contained in the Apollo parachute system. The reliability requirement of independent parachute deployment, coupled with large command module oscillations, necessitates divergent drogue parachute and main

---

67. NASA Manned Spacecraft Center, "NASA Support Manual: Apollo Spacecraft Familiarization," SM2A-02 (December 1, 1966), pp. 3-8, 3-9.

68. NASA Manned Spacecraft Center, "Project Apollo Spacecraft Description," NASA TM-X-619452 (September 1, 1963), p. 3.114.

The Apollo 16 Command Module, with astronauts John W. Young, Thomas K. Mattingly II, and Charles M. Duke, Jr., aboard, nears splashdown in the central Pacific Ocean to successfully conclude a lunar landing mission. This overhead picture was taken from a recovery helicopter seconds before the spacecraft hit the water on April 27, 1972, with its three parachutes allowing a safe return of the capsule and crew near Christmas Island in the Pacific. NASA S72-36287.

pilot parachute deployment angles coupled with positive thruster type deployment. The command module oscillations create the possibility of contact between the parachute risers and the hot rear heat shield, and last but not least, the increase in CM weight without an accompanying increase in compartment volume or allowable parachute cluster loads resulted in novel design approaches for parachute packing, storage and shape retention.

Designed for use in both optimum and crisis situations, either during launch abort or return from the Moon, this system was fully redundant and handled forces equivalent to 3 g's without difficulty. A study explains that as designed, the system should do the following:

Two ribbon drogue parachutes accomplish initial deceleration and stabilization, with only one parachute being required and the second parachute providing the back-up mode. Deploying both parachutes simultaneously eliminates the need for an emergency sensor, provides for faster CM stabilization and creates more favorable main parachute deployment conditions. After disconnect the two drogue parachutes are followed by three pilot parachute deployed Ringsail main parachutes; two of which will provide the rate of descent necessary for water landing.

The study concluded the following: "The basic design proved flexible enough to accept a substantial increase in command module weight and a resultant increase in recovery envelope and velocity of parachute deployment without changing parachute volume or load requirements."[69]

Not all went well with every aspect of the Apollo parachute-recovery system. A number of tests failed during the runup to the missions to the Moon. For example, on September 6, 1963, Apollo command-module boilerplate no. 3 was destroyed when one pilot parachute was cut by contact with the vehicle and one of its main parachutes did not deploy. Then, rigging problems caused the other two parachutes to fail. An investigation led to rigging and design changes on future systems. In this case, these difficulties were resolved and the program continued.[70] The most serious failure came during the descent of Apollo 15 from the Moon in 1971. During its reentry, all three main parachutes deployed without incident at an altitude of 10,000 feet, but one of the three parachutes deflated while the Apollo 15 capsule was obscured by clouds between 7,000 and 6,000 feet. Regardless, the crew and spacecraft returned safely because redundancy in the system allowed one parachute to fail without compromising the entire landing system. However, short the one parachute, the crew descended to impact at a slightly higher velocity than planned. Failure analysis found that the parachute lines had been damaged by fuel from the reaction control system (RCS) during return, a normal occurrence, but in this instance, the parachute assembly was in the way of the RCS ejection ports. The Apollo mission summary reported only: "During the descent, one of the three main parachutes failed, but a safe landing was made."[71] As reported at the time: "The most probable cause of the anomaly was the burning of raw fuel (monomethyl hydrazine) being expelled during the latter portion of the depletion firing and this resulted in

---

69. T.W. Knacke, Northrop Ventura, "The Apollo Parachute Landing System," paper presented at the AIAA Second Aerodynamic Decelerator Systems Conference, El Centro, CA, September 1968.

70. Mark Wade, "Apollo Parachute Landing System," *http://www.astronautix.com/craft/csmchute.htm*, accessed September 12, 2009.

71. NASA Johnson Space Center, "Apollo Program Summary Report," JSC-09423 (April 1975), pp. 2–48.

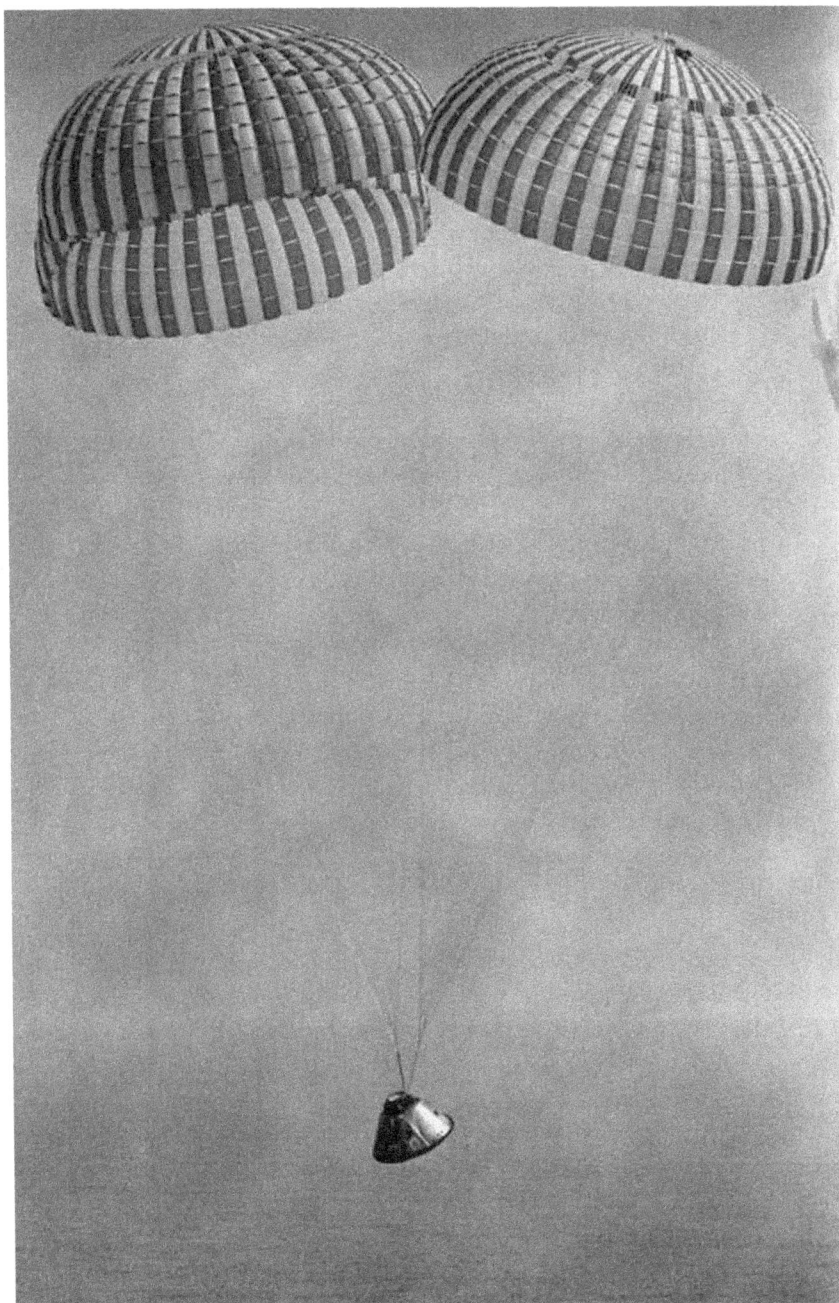

The Apollo 9 spacecraft, with astronauts James A. McDivitt, David R. Scott, and Russell L. Schweickart aboard, approaches splashdown in the Atlantic recovery area on March 13, 1969, only 4.5 nautical miles from the prime recovery ship, USS *Guadalcanal* (LPH-7). NASA S69-27467.

exceeding the parachute-riser and suspension-line temperature limits."[72] Based on this anomaly and its occurrence frequency—only once in all of the missions to date—NASA investigators believed that there was only a 1-in-17,000 chance of failure on future missions.[73]

This basic approach worked well throughout Apollo, but earlier NASA engineers still wanted to have a nonwater landing system similar to that pursued for Gemini. Personnel at the Landing Technology Branch of NASA's Manned Spacecraft Center tried to adapt the parasail landing system underway for Gemini to the Apollo Application Program, the follow-on to the Moon landings. The branch reported that it

> expects to have a system that will be adaptable to Apollo. Their present effort is not aimed directly at incorporation of such a system, but rather at developing the technology and hardware necessary for the system itself. They are, however, basing their designs on a spacecraft that is of the CM size and type. The resulting system will most likely consist of a steerable parachute, plus some combination of landing rockets, deployable energy absorbers and stability aids.[74]

Its staff added that it had contracted with four organizations for various aspects of this effort:

A. Bendix Products Aerospace Division is developing a computer program to analyze land-landing dynamics.
B. North American Aviation is investigating landing gear systems for the Command Module.
C. Pioneer Parachute is developing a parasail type of steerable parachute.
D. Northrop Ventura is developing a cloverleaf type of steerable parachute.

The primary concern was whether this landing system could handle a 14,000-pound capsule and be containable within a 1.5-cubic-meter space. While this system was considered even less heavy and bulky than a water system, the addition of landing rockets to cushion a landing might push the total weight above what was already envisioned for the Apollo Command Module. "Probable weight increase and cost of incorporation must be weighed against the added capability and decrease in cost of recovery operations," the

---

72. NASA Manned Spacecraft Center, "Apollo 15 Main Parachute Failure Anomaly Report No. 1," NASA TM-X-6835 (1971).
73. NASA Manned Spacecraft Center, "Apollo Main 15 Parachute Failure," n.d.
74. W.W. Hough, "Apollo Application Program Land-Landing System—Case 218, Task 18," NASA-CR-15644, Bellcomm, Inc. (December 13, 1965).

The Apollo 13 Command Module splashes down, and its three main parachutes collapse, as the week-long, trouble-plagued Apollo 13 mission comes to a premature but safe end. The spacecraft, with astronauts James A. Lovell, Jr., commander; John L. Swigert, Jr., Command Module pilot; and Fred W. Haise, Jr., Lunar Module pilot aboard, splashed down on April 17, 1970, in the South Pacific Ocean, only about 4 miles from the USS *Iwo Jima* (LPH-2) prime recovery ship. NASA S70-35644.

study concluded.[75] This was the first reference in this recovery literature from the 1960s concerning the significant tradeoff engineers had to make between added weight and reduced-recovery operational cost. The Navy was generally quite agreeable during the space race era to deploying its ships for recovery, and NASA was not required to pay for that total operation. That made water recovery, at least from NASA's perspective, not only the most expedient option, but also the least expensive method of recovery. Even so, this program concluded without adopting anything more sophisticated than the parachute system used for Mercury, Gemini, and Apollo. NASA would not return to the parasail/paraglider concept for landing until the 1990s.

---

75. Hough, "Apollo Application Program Land-Landing System—Case 218, Task 18"; D.G. Estberg, Bellcomm Inc., "The Probability of Land Landing after an LEV Abort," TM-68-2013-6, CR-100283 (December 31, 1968).

DON'T BE RESCUED FROM

OUTER SPACE

FLY BACK IN STYLE

Wen Painter of NASA's Flight Research Center drew this cartoon in 1966. It captured what some saw as a key difference between space capsule splashdowns at sea and spaceplane landings on a runway. NASA.

CHAPTER 4

Spaceplane Fantasies

While NASA's space capsules presented a near-term capability for human space flight, few in the industry believed the vehicles were ideal. The future, as envisioned by almost everybody, included a lifting-reentry spaceplane that would land on a conventional runway. Even the capsule designers recognized the advantages of a lifting reentry vehicle, and Gemini and Apollo each had a small L/D ratio to allow more precise targeting of a landing site.[1]

During a lifting reentry, a vehicle generates lift perpendicular to its flight-path, which can then be adjusted to change both vertical motion and flight direction. Practical entry angles have an upper and lower limit. The lower limit, also called the overshoot boundary, is the angle at which the vehicle will skip back out of the atmosphere (essentially the normal angle proposed for skip-gliders). The upper limit, or undershoot boundary, is the load-factor limit established by vehicle-structural, human-tolerance, or operational considerations. The primary measure for lifting reentry is the hypersonic L/D coefficient.[2] Increasing the L/D has significant effects on decreasing the maximum entry load and increasing the allowable entry corridor depth. Low L/D values (0.5–1.0) produce durations essentially the same as a semiballistic reentry, with survivable g-loads, moderate heating levels, and low maneuverability.

---

1. Some purists might take exception to our use of the term "spaceplane" to characterize winged/lifting body reusable space vehicles. The reality, however, is that this term describes any "aircraft that takes off and lands conventionally but is capable of entry into orbit or travel through space" (*Oxford Dictionary*, available online at *http://oxforddictionaries.com/definition/spaceplane*, accessed June 7, 2011). Similarly, one definition is "A rocket plane designed to pass the edge of space, combining certain features of aircraft and spacecraft" (*http://www.wordnik.com/words/spaceplane*, accessed June 7, 2011).

2. After about 1965, discussions of hypersonic L/D generally assumed the definition used by Al Draper at the Air Force Flight Dynamics Laboratory, which measured the maximum L/D at 20,000 feet per second at an altitude of 200,000 feet. This definition provided a consistent basis for comparison and was significant since hypersonic viscous effects could have a large effect on the maximum L/D.

Increasing the hypersonic L/D above 1.0 brings only marginal improvements in entry loads, but extends the cross range as a nonlinear function of the L/D (an L/D of 1.5 might generate a 1,000-mile cross range, but an L/D of 3 will generate a 4,000-mile cross range). High L/D values produce low g-loads and allow significant variations in the flightpath, but they result in long-duration reentries with continuous heating. Although the peak temperatures of a lifting reentry are lower than those of a ballistic reentry, the total heat load (temperature multiplied by duration) is generally higher.[3]

Researchers expected that a spaceplane would land on a preselected runway, like an airplane, and would not require a large recovery force, like capsules. Theoretical and wind tunnel studies, however, showed the aerodynamic configurations that produced the highest L/D during entry did not necessarily produce a high L/D at landing speeds. Long slender cones or wedges without wings were ideal for entry, but the best landing configurations used long, gliderlike wings. Engineers soon discovered that by providing two flight modes—one for hypersonic velocities through the transition to subsonic flight, and another for landing—many of the flight-control problems of a spaceplane could be decoupled.

Since the majority entry occurs at hypersonic speeds, the engineers decided to concentrate on developing shapes that exhibited adequate stability and control at high speeds, and then they would add the equipment for the decoupled landing mode. During the late 1950s and early 1960s, the more prominent landing techniques (see chapter 2) included gliding parachutes, paragliders (Rogallo wings), rotors, sustained propulsive lift, and deployable wings (usually, incorrectly, called variable-geometry wings). However, it quickly became evident that this type of equipment greatly expanded the complexity of the vehicle and added a significant amount of weight. Researchers eventually found that the delta wing provided a compromise that allowed a single shape to perform well at hypersonic velocities and adequately during landing.

Unfortunately, thermal protection systems for a spaceplane were a great deal more challenging than for capsules due to the higher total heat load generated by the longer entry time. At the same time, the large surface of the delta wing further exacerbated matters. Needing to maintain a smooth aerodynamic surface added yet another complication.

In spite of any perceived drawbacks, lifting reentry spaceplanes capable of landing on a normal runway were considered necessary to increase human access to space at a reasonable cost, and became the Holy Grail of the aerospace

---

3. *Testing Lifting Bodies at Edwards*; Christopher J. Cohan, Lifting Hypersonic Vehicles, a Short Course on Technology of Space Shuttle Vehicles, the University of Tennessee Space Institute, Tullahoma, TN, October 1970.

industry. At the time, NASA officials compared the space program's traditional use of expendable launch vehicles with capsules atop to throwing away a railroad locomotive after every train trip, whereas a reusable spaceplane would offer cost-effective, routine access to space.[4]

Choices

In the immediate postwar era, three basic vehicle types were being considered for high-speed global flight: (1) ballistic, (2) boost-glide, and (3) skip-glide. Early in 1954, H. Julian Allen, Alfred J. Eggers, Jr., and Stanford E. Neice at the NACA Ames Aeronautical Laboratory put together a theoretical discussion of these different concepts in *A Comparative Analysis of the Performance of Long-Range Hypervelocity Vehicles.* In that report, the researchers examined the relative advantages and disadvantages of each vehicle type. The conclusions proved significant. They found that ballistic and boost-glide vehicles could probably be developed, with proper attention to design. However, even at this early date, the researchers warned that skip-gliders appear substantially less promising.[5]

The flight profile of each vehicle type is subtly different. The ballistic vehicle, best exemplified by the long-range missile, leaves the atmosphere and returns as a single maneuver, generally in the form of a parabola. A boost-glider is accelerated to a speed and altitude where the dynamic pressure allows the vehicle to glide at some given L/D coefficient. At this altitude, termed by Eugen Sänger the "equilibrium altitude," the aerodynamic lift required for flight is less than the centrifugal force resulting from the curved flight around Earth.[6]

---

4. Roger D. Launius, "NASA and the Decision to Build the Space Shuttle, 1969–72," *The Historian* 57 (Autumn 1994): 17–34.

5. Alfred J. Eggers, Jr., H. Julian Allen, and Stanford E. Neice, "A Comparative Analysis of the Performance of Long-Range Hypervelocity Vehicles," NACA report TR-1382. A modified version of this paper was published by Allen as "Hypersonic Flight and the Reentry Problem," *Journal of the Aeronautical Sciences* 25 (April 1958): 217–230; H. Julian Allen, "Hypersonic Flight and the Reentry Problem," *Twenty-First Wright Brothers Lecture*, December 17, 1957, later published as NASA-TM-108690.

6. Eugen Sänger, *Raketenflugtechnik* (Berlin, Germany: Oldenbourg, 1933), p. 112, later translated by NASA as TT F-223 (1965); Eugen Sänger and Irene Bredt, *A Rocket Drive for Long-Range Bombers,* from the German *Über einen Raketenantrieb für Fernbomber,* translated by the Naval Technical Information Brach, Bureau of Aeronautics as CGD-32 (Washington, DC: U.S. Navy, 1952), p. 60.

The required aerodynamic lift lessens as speed increases, and it becomes zero at orbital velocities. Surprisingly, the equilibrium altitude is fairly low—usually below 250,000 feet—until the speed approaches orbital velocity. Once at the equilibrium altitude, the vehicle maintains its maximum L/D until aerodynamic drag causes it to slow and lose altitude until it lands.[7]

On the other hand, the trajectory of a skip-glider (such as the Sänger Silverbird) is composed of a succession of ballistic paths, each connected to the next by a skipping phase during which the vehicle enters the atmosphere, negotiates a turn at some given L/D coefficient, and is ejected from the atmosphere by aerodynamic lift. Each upward skip results in a lower peak altitude than the last since the vehicle is unpowered, and this sequence continues until the vehicle no longer has sufficient energy to leave the sensible atmosphere, at which time it glides to a landing.[8]

While he was investigating the relative merits of the various vehicle concepts, Allen compared the likely structural weights of each vehicle. He noted that the apparent advantage of the ballistic missile—a lack of wings—was largely negated by the need for increased propellant tankage since the vehicle could not rely on aerodynamic lift at any point in its trajectory. The ballistic vehicle also experienced the highest aerodynamic loads while entering the atmosphere, adversely affecting structural weight (although the skip-glider suffered a similar flaw). In this regard, a boost-glider, or spaceplane, was the most efficient configuration in terms of structural weight, assuming a suitable, lightweight thermal protection system could be found.[9]

The researchers noted that the three vehicle types reacted to entry heating in vastly different ways. The blunt-body theory made it possible to design a ballistic reentry vehicle because the majority of the heat was carried away from the vehicle by shock waves and other aerodynamic phenomena; heat sinks or ablators could handle the residual heat. Obviously, this was impractical for both of the glide vehicles since, by definition, a glider needs a high L/D to increase its range and avoid plunging through the atmosphere.[10]

However, a boost-glider could gradually convert its kinetic energy over a longer time than a capsule, radiating much of the heat back into the atmosphere

---

7. Allen, "Hypersonic Flight and the Reentry Problem," p. 6.

8. Ibid.; Alvin Seiff and H. Julian Allen, "Some Aspects of the Design of Hypersonic Boost-Glide Aircraft," NACA report RM A55E26 (August 15, 1955), pp. 1–2; H. Julian Allen and Stanford E. Neice, "Problems of Performance and Heating of Hypersonic Vehicle," NACA RM A55L15 (March 5, 1956), pp. 1–2.

9. Allen, "Hypersonic Flight and the Reentry Problem," pp. 12–13.

10. Ibid., pp. 13–14.

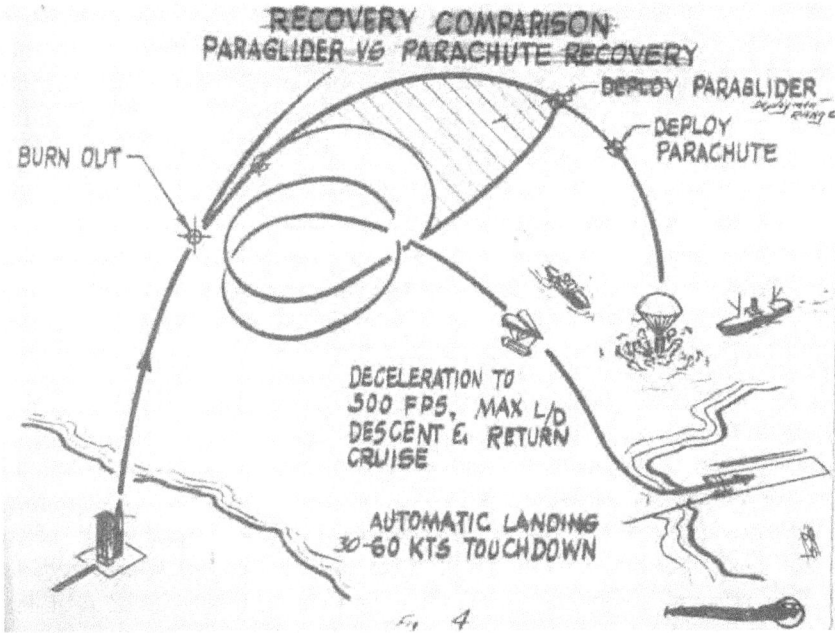

This early 1960s comparison of the relative merits of a parachute landing system versus a winged approach clearly favored flying back to a runway. NASA.

and maintaining comparatively low structural temperatures. Testing by Allen and Eggers revealed that with sufficiently low wing loading, it might be possible to build a glider that could radiate enough heat to maintain a structural temperature under 1,600 °F, which was within the capability of some available materials. The wing loading had to be kept low since increasing it adversely affected the radiation-equilibrium temperature, resulting in large wings and more thermal protection. Active cooling systems could be employed at the hot spots (nose, wing leading edge, etc.), where the equilibrium temperature exceeded the tolerance of available materials.[11]

The skip-glider, on the other hand, seemed to be unworkable. A large fraction of the kinetic energy was converted to heat in a short time during each skip, but the interval between skips was not sufficient to radiate the heat into space.[12]

---

11. Ibid., pp. 14–15.

12. Radiative cooling was most effective through radiating into space (that is, out of the top of the vehicle). Radiating back toward Earth was fairly ineffective since the atmosphere is already at fairly high temperatures compared to the cold of space. X-20 radiated from the lower skin across the structure to the upper skin, which then radiated to space.

Eggers determined that any skip-glider would need to use an extensive active cooling system, and the weight penalty would probably be so excessive as to rule this vehicle out as impractical, or even impossible, except for very short flights.

Despite having discovered the blunt-body theory that made ballistic reentry possible, Eggers became convinced of the overall desirability of a lifting reentry vehicle instead of a ballistic capsule. He revealed this preference at the June 1957 annual meeting of the American Rocket Society in San Francisco, CA. Eggers was skeptical about the relatively high heating loads and deceleration forces characteristic of ballistic reentry, warning that "the g's are sufficiently high to require that extreme care be given to the support of an occupant of a ballistic vehicle during atmospheric reentry." He also pointed out that such an object, entering the atmosphere along a shallow trajectory to hold deceleration down to 7.5 g's, would generate a surface temperature of at least 2,500 °F. Thus, in Eggers's judgment, "the glide vehicle is generally better suited than the ballistic vehicle for manned flight."[13] He also saw the difficulty of recovering a ballistic capsule since it was not controllable in the atmosphere and might need a target area of several thousand square miles. Of course, several of these challenges were subsequently overcome by providing a small amount of lift, and developing better ablators, for the capsules.

Many concepts for accomplishing a lifting reentry followed by a horizontal landing were proposed during the late 1950s. There were inflatable vehicles with very low wing loading; delta-winged flattop shapes; delta-winged flat-bottom shapes; semiballistic shapes with extendable wings; and several lifting bodies. A team at NACA Langley that included Charles H. McLellan, who had solved many of the aero-thermal issues associated with the X-15, determined in 1959, "in many ways body shape is of secondary importance to other design parameters such as size, weight, planform loading, and lift-drag ratio. Furthermore, for a given lift-drag ratio and weight-to-base-area ratio, no specific body shape was found to possess sufficiently superior qualities to exclude consideration of other body shapes. However, the configurations with more nearly flat bottoms appeared to offer some advantages from the standpoint of performance and heat-transfer considerations."[14]

---

13. Alfred J. Eggers, "Performance of Long Range Hypervelocity Vehicles," *Jet Propulsion* 27 (November 1957): 1,147–1,151. The peak temperatures on the heat shield of the Mercury spacecraft during its reentry from orbit reached approximately 3,000 °F.

14. William O. Armstrong et al., "The Aerodynamic-Force and Heat-Transfer Characteristics of Lifting Reentry Bodies," a paper presented at the Joint Conference on Lifting Manned Hypervelocity and Reentry Vehicles on the Langley Research Center, April 11–14, 1960, pp. 141–142.

Lifting Bodies

Other researchers realized that the blunt-body shape could be modified into a lifting body that would generate a modest 1.5 hypersonic L/D and enable potential landing sites to be several hundred miles cross range. Even this relatively low L/D would be a great improvement over the capsules: Mercury had essentially zero L/D, but the slight refinements to Gemini raised this to about 0.25 at hypersonic velocities. Apollo is generally credited with an L/D between 0.6 and 0.8, but the original specifications only required 0.4 to provide adequate margins for guidance errors during super-orbital reentries (i.e., direct return from the Moon). However, the guidance techniques actually used during Apollo were sufficient. Therefore, most of the L/D was not required, and the flown missions used an L/D between 0.29 and 0.31, subjecting the crew to a deceleration of about 6 g's.[15]

Apparently, Sänger had first proposed the concept of a lifting body as early as 1933. Sänger argued that a flat-bottom lifting body would have a higher L/D at high supersonic speeds than a normal-shape aircraft of similar size.[16] However, he did not pursue this concept to the extent of determining the optimum shape for a lifting body, nor did he prove that such a vehicle should necessarily have a flat bottom—although all of the Silverbird designs did feature flat bottoms.

In 1954, Meyer M. Resnikoff at NACA Ames began looking at the earlier work by Sänger and determined that a lifting body would enjoy a 40- to 100-percent advantage in L/D over a conventional aircraft at hypersonic velocities. Lacking access to a hypersonic wind tunnel, the study was based instead on formulas derived from Sir Isaac Newton's impact theory.[17] Resnikoff concluded that "the lower surface of such a body must be flat [thus verifying Sänger's speculation] and rectangular, and that if the maximum available

15. E.P. Smith, "Space Shuttle in Perspective: History in the Making," AIAA Paper 75-336, presented at the AIAA 11th Annual Meeting and Technical Display, Washington, DC, February 24–26, 1975, p. 4.

16. See, for instance, Eugen Sänger, *Raketenflugtechnik* (Berlin: Oldenbourg, 1933), pp. 112, 120–121, later translated by NASA as TT F-223 (Washington, DC: NASA, 1965); Eugen Sänger and Irene Bredt, *A Rocket Drive for Long-Range Bombers*, from the German *Über einen Raketenantrieb für Fernbomber*, translated by the Naval Technical Information Brach, Bureau of Aeronautics as CGD-32 (Washington, DC: U.S. Navy, 1952), pp. 58–64.

17. Isaac Newton, *Principia* (first published in 1687) as translated by Andrew Motte in 1729, revised by the University of California in 1946, pp. 333, 657–661.

volume is utilized, the minimum drag body…is a wedge."[18] It should be noted that the lifting bodies investigated by Resnikoff bore little resemblance to most of the vehicles ultimately flown during the 1960s. Other researchers, such as Dr. Wilbur L. Hankey, Jr., at the Air Force Flight Dynamics Laboratory, reached similar conclusions around the same time.[19]

A different lifting body concept was presented in a paper presented by Ames's Thomas J. Wong at the same March 1958 conference at which Maxime A. Faget detailed his blunt-body capsule design. Wong showed a 30-degree half-cone with an L/D high enough to limit deceleration during reentry to approximately 2 g's and to allow a lateral cross range of ±230 miles, with a longitudinal variation of 700 miles. Unfortunately, much like the winged-craft proposal presented by NACA Langley Research Center's John V. Becker at the same conference, this vehicle weighed substantially more than a pure blunt-body shape, and the throw-weight of the Redstone and Atlas boosters being used for Project Mercury did not allow such luxuries.[20]

Nevertheless, the lifting body concept interested the Air Force, which subsequently funded several small research programs, including one with The Aerospace Corporation that would enjoy sustained support over many years. The Air Force was interested in a maneuverable intercontinental ballistic missile (ICBM) warhead that could reduce the effectiveness of any antiballistic missile system, and the ability to select a landing site away from the ground track of a photoreconnaissance satellite could allow more opportunities for a film-return vehicle to be recovered. This was particularly important to polar operations, where the natural rotation of Earth moves the launch site considerably cross range during even a short orbital flight. High cross range allows a vehicle to land at its launch site after a single orbit, a desired capability that would later drive much of the Space Shuttle configuration. Ultimately, the warhead community moved in an entirely different direction with the high-ballistic coefficient reentry vehicles.

There were other perceived benefits of the lifting body design. It appeared that for any given planform size (wing area), the lifting body offered a large internal volume compared to a capsule or wing-body shape. However, this would prove misleading because much of the volume was composed of areas

18. Meyer M. Resnikoff, "Optimum Lifting Bodies at High Supersonic Airspeeds," NASA Report RM A54B15 (May 7, 1954), pp. 1, 12.

19. See, for instance, Roland N. Bell, "A Closed-Form Solution to Lifting Reentry," Air Force Flight Dynamics Laboratory (AFFDL) report 65-065 (1965).

20. Thomas J. Wong et al., "Preliminary Studies of Manned Satellites—Wingless Configurations: Lifting-Body," NACA Conference on High-Speed Aerodynamics, pp. 35–44.

Three lifting body vehicles at NASA's Flight Research Center in California in 1969. Left to right: the X-24A, M2-F3, and HL-10. These aircraft helped pave the way for later spaceplane concepts. NASA EC69-2358.

with complex curves that did not lend themselves to being efficiently filled with normally square (cube) equipment. Even propellant tanks presented a challenge since they needed to be odd shapes—often a difficult proposition with cryogenic or high-pressure storage vessels. Ultimately, packaging equipment effectively into a lifting body proved challenging both conceptually and practically.

By the end of 1958, four very different lifting body shapes were being seriously studied, including the Ames M1, the Langley HL-10, the Langley lenticular vehicle, and The Aerospace Corporation's A3. Several other shapes were also under evaluation at the Air Force Flight Dynamics Laboratory (AFFDL) at Wright-Patterson AFB in Dayton, OH.[21] Each represented a distinctive design approach.

Conceived largely by Eggers at NACA Ames, the M1 began as a modified 13-degree half-cone, mostly flat on the top, with a rounded nose to reduce heating. The shape had a hypersonic L/D of 0.5 (only marginally better than a ballistic blunt-body capsule) and demonstrated a pronounced tendency to tumble end over end at subsonic speeds. In fact, the shape had virtually no subsonic L/D and could not land horizontally. Eventually, the Ames researchers Eggers, Clarence A. Syvertson, George G. Edwards, and George C. Kenyon

---

21. The Air Force Flight Dynamics Laboratory was established on March 8, 1963, at Wright-Patterson AFB from a reorganization of the earlier Directorate of Aeromechanics.

discovered that most of the stability problems could be cured by modifying the aft end with body flaps that looked much like a badminton shuttlecock, creating the M1-L. The slightly improved L/D would give the vehicle about 200 miles of lateral maneuverability during reentry and about 800 miles of longitudinal discretion over its landing point.[22]

This design gradually morphed into a blunted 26-degree cone that offered improved stability, and it became the M2 shape. Further flattening of the cone top provided additional lift and a hypersonic L/D of 1.4. The base area was reduced by adding a boat tail (first proposed in 1960 by Edwards and David H. Dennis at Ames) to the upper and lower surfaces of the half-cone, simultaneously improving the reentry trim capability and reducing drag. Soon, a protruding canopy, twin vertical stabilizers, and various control surfaces were added, leading to it being called the M2b "Cadillac." Preliminary wind tunnel tests showed that this configuration would have a subsonic L/D of about 3.5 and could probably be landed safely if an adequate control system could be developed.[23]

At NACA Langley, a different shape was conceived under the general guidance of Eugene S. Love, John W. Paulson, Jr., and Robert W. Rainey. It was based on studies initiated in 1957 that showed a flat-bottom reentry configuration with negative camber (cross section like an inverted-wing airfoil) would have a 14-percent higher L/D than a blunt half-cone. Sharply upswept tips provided control with a single centrally mounted vertical stabilizer. Langley

---

22. George H. Holdaway, Joseph H. Kemp, Jr., and Thomas E. Polek, "Aerodynamic Characteristics of a Blunt, Half-Cone Entry Configuration, with Horizontal Landing Capability," NASA TM-X-1029 (October 1, 1964); Holdaway, Kemp, and Polek, "Characteristics of an Aerodynamic Control System for use with Blunt Entry Configurations at Mach Numbers of 10.7 and 21.2 in Helium," NASA TM-X-1153 (October 11, 1965); John L. Vitelli and Richard P. Hallion, "Project PRIME: Hypersonic Reentry From Space," *Case V of The Hypersonic Revolution: Case Studies in the History of Hypersonic Technology*, Volume 1, From Max Valier to Project PRIME (1924–1967) (Bolling AFB, DC: USAF Histories and Museums Program, 1998), p. 529; "Piloted Lifting-Body Demonstrators," p. 866; Richard P. Hallion, *On the Frontier: Flight Research at Dryden, 1946–1981* (Washington, DC: NASA SP-4303, 1984), p. 148.

23. Robert G. Hoey, *Testing Lifting Bodies at Edwards* (Edwards AFB, CA: PAT Projects, 1994); "Backup Program for Modified 698N Configuration," Martin report ER 12147, April 1962, pp. 1–2; "Project PRIME: Hypersonic Reentry," pp. 535–536; Clarence A. Syvertson, "Aircraft Without Wings," *Science Journal*, December 1968; David H. Dennis and George G. Edwards, "The Aerodynamic Characteristics of Some Lifting Bodies," pp. 103–104, a paper presented at the Joint Conference on Lifting Manned Hypervelocity and Reentry Vehicles, Langley Research Center, April 11–14, 1960.

continued developing this concept and, in 1962, the configuration was designated HL-10 (HL stands for horizontal lander).[24]

Researchers at Langley also proposed a lenticular shape. Donna Reed, the wife of Robert Dale Reed, who will play a major role in this story a bit later, called this a powder puff; others called it a flying saucer. The design was a circular shape that had no noticeable protuberances during entry. It transitioned to horizontal flight by extending control surfaces after following a reentry profile similar to a symmetrical capsule. Although various artists' concepts show the shape, and it is briefly mentioned in several technical reports, the only serious investigation of it appears to have been during the development of the Convair Pye Wacket defensive missile for the B-70 (WS-110A) bomber.[25]

A team led by Frederick Raymes at The Aerospace Corporation developed the original A3 shape. The design was subsequently transferred to the Glenn L. Martin Company, where engineers, under the direction of Hans Multhopp, found they could improve its L/D at both hypersonic and subsonic speeds by using a more slender body and a flatter lower surface. After consulting with engineers at the AFFDL, the Martin engineers introduced positive camber in the longitudinal axis (the opposite of that used on the HL-10). John Rickey, an aerodynamicist at Martin, was assigned to refine the configuration into a practical design as part of Air Force Project M-103. The resulting A3-4 or SV-5 (space vehicle-5) shape had a severe delta planform with pronounced rounding and twin vertical stabilizers. It appears that the shape was specifically tailored to meet anticipated Air Force cross-range requirements for the CORONA film-return vehicles.[26]

Wind tunnel tests conducted on the SV-5 shape led to several changes to enhance the transonic characteristics, including slightly altering the angle of

---

24. John W. Paulson, Jr., "Low-Speed Static Stability Characteristics of Two Configurations Suitable for Lifting Reentry from Satellite Orbit," NASA MEMO-10-22-58L (November 1958); Robert W. Kempel, Weneth D. Painter, and Milton O. Thompson, "Developing and Flight Testing the HL-10 Lifting-Body: A Precursor to the Space Shuttle," NASA Reference Publication 1332 (April 1, 1994).

25. R. Dale Reed with Darlene Lister, *Wingless Flight: The Lifting-Body Story* (Washington, DC: NASA SP-4220, 1997), pp. 14–15. For a further description of the unusual Pye Wacket concept, see Dennis R. Jenkins and Tony R. Landis, *Valkyrie: North American's Mach 3 Superbomber* (North Branch, MN: Specialty Press, 2003).

26. SAMOS Satellite Reconnaissance System, "Briefing to the Satellite Intelligence Requirements Committee," undated (but in the March–May 1960 timeframe); "Project PRIME: Hypersonic Reentry," pp. 538–539; Correspondence between Frederick Raymes and Dennis R. Jenkins during 1992–1996; US Patent 203,902, August 23, 1963. The Martin contract was AF-4(695)-103.

NASA engineer Dale Reed holds a model of the M2-F1 lifting body with the full-scale version directly behind him in 1967. In support of the M2 lifting body program in the early 1960s, Reed had built a number of small lifting body shapes and drop-tested them from a radio-controlled mothership. NASA E-16475.

the nose ramp to provide better high- and low-speed trim capability. As finalized in late 1963, the configuration incorporated both a center-fin and tip-fin airfoil that included inward camber.[27]

The concept of lifting reentry was taken seriously enough that a Joint Air Force–NASA Conference on Lifting Manned Hypervelocity and Reentry Vehicles was held at NASA Langley on April 11 to 14, 1960. The goal was to disseminate research results relating to piloted hypervelocity and reentry vehicles, which have lifting capability and the ability to maneuver in the atmosphere. Papers were presented that covered many aspects of vehicle design, wind tunnel and theoretical testing, flutter, guidance and control, heat transfer, and other issues of interest to vehicle designers. The results from hypersonic wind tunnel and hypervelocity-facility tests allowed researchers to predict that each of the lifting bodies had acceptable hypersonic characteristics.[28]

Air Force–Sponsored Research

In 1959, the Air Force evaluated the high-speed characteristics of a lifting body during three test flights that were part of a larger research effort called WS-199. With significant involvement from the AFFDL, McDonnell developed the Model 122B Alpha Draco boost-glider as part of WS-199D. Each lifting body–shaped glider was launched to 92,000 feet by a two-stage launch vehicle derived from the Sergeant Battlefield missile. After being accelerated to Mach 5, the Alpha Draco vehicle glided toward a preprogrammed location about 240 miles downrange, where it entered a terminal dive into the ocean. The first two Alpha Draco flights on February 16 and March 16, 1959, were successful, but the third flight, on April 27, 1959, did not fare as well and was destroyed by range safety after it deviated from its planned flightpath. Despite the failure, the program verified the basic principles of boost-glide vehicles, and provided limited data on hypersonic aerodynamics and thermodynamics.[29]

By August 1959, engineers at the AFFDL believed the advent of new high-temperature materials and sophisticated guidance packages would permit

---

27. *Testing Lifting Bodies at Edwards.*

28. The papers were subsequently published in *A Compilation of Papers Presented at the Joint Conference on Lifting Manned Hypervelocity and Reentry Vehicles on 11–14 April 1960 at the Langley Research Center,* NASA classified report TM-X-67563.

29. Alfred C. Draper and Thomas R. Sieron, "Evolution and Development of Hypersonic Configurations, 1958–1990," Air Force report WL-TR-91-3067 (September 1991), pp. 1–2; Joel W. Powell, "DRACO: The 'Secret' Launches at Cape Canaveral in 1959," *Spaceflight,* April 2000.

testing hypersonic gliders during actual reentry flights. The laboratory originally envisioned the gliders as simple wing-body vehicles with a pronounced
keel on the ventral surface. However, Alfred C. Draper, Jr., recognized that
the keel design would introduce a dihedral effect and result in serious heating
problems during reentry. Draper successfully argued that the gliders should
instead use the WLB-1 shape that had recently been developed by the laboratory. Ultimately, however, the final configuration was essentially the forward 4
feet of the Dyna-Soar glider, with a flat-bottom delta planform, rounded nose
cap, rounded wing leading edges, and a tilted nose for hypersonic trim. The
leading edges, lower surfaces, and upper surfaces were of the same design and
materials as Dyna-Soar, allowing designers to take advantage of the large body
of wind and shock tunnel data that was being accumulated by Dyna-Soar.[30]

On January 31, 1961, the effort was designated Project 1466 and named
Aerothermodynamic/Elastic Structural Systems Environmental Tests (ASSET).
The flight program was designed to assess the applicability and accuracy of
analytical methods and experimental techniques in the areas of structures,
aerodynamics, aerothermodynamics, and aerothermoelasticity for lifting reentry vehicles. In April 1961, the Air Force awarded a contract to McDonnell
for two slightly different glider configurations: aerothermodynamic structural
vehicle (ASV) and aerothermoelastic vehicle (AEV). McDonnell built four
of the former and two of the latter. The gliders were 5.7 feet long, weighed
1,130 pounds (ASV) and 1,225 pounds (AEV), and had a sharply swept (72.5
degrees) low-aspect-ratio delta wing. The anticipated Blue Scout boosters were
soon replaced by Thor launch vehicles, which were more readily available.[31]

Although superficially similar, and sharing common subsystems, the two
types of gliders differed completely in mission and research capabilities, and

---

30. "History of the Aeronautical Systems Division, January Through June 1962," Aeronautical
Systems Division report, December 1962; Richard P. Hallion, "ASSET: Pioneer of Lifting Reentry,"
Case IV of *The Hypersonic Revolution: Case Studies in the History of Hypersonic Technology,
Volume 1, From Max Valier to Project PRIME (1924–1967)* (Bolling AFB, DC: USAF Histories and
Museums Program, 1998), pp. 449–452; *Testing Lifting Bodies at Edwards.*

31. "ASSET," pp. 1–2; "Advanced Technology Program: Technical Development Plan for Aerothermodynamic/
Elastic Structural Systems Environment Tests (ASSET)," AFFDL report, 1963, pp. 21–22, 43–44; *ASSET:
Pioneer of Lifting Reentry,* pp. 449–452. The McDonnell contract was AF33(616)-8106; Alfred C. Draper
and Thomas R. Sieron, "Evolution and Development of Hypersonic Configurations, 1958–1990," Air
Force report WL-TR-91-3067, September 1991, pp. 9, 15. These Thor intermediate-range ballistic mis-
siles were part of a group returned from the United Kingdom after having been deployed for several years
and were available at extremely reasonable costs. Contracting problems with NASA (who managed the
Delta upper stage) resulted in the first ASV being launched without it.

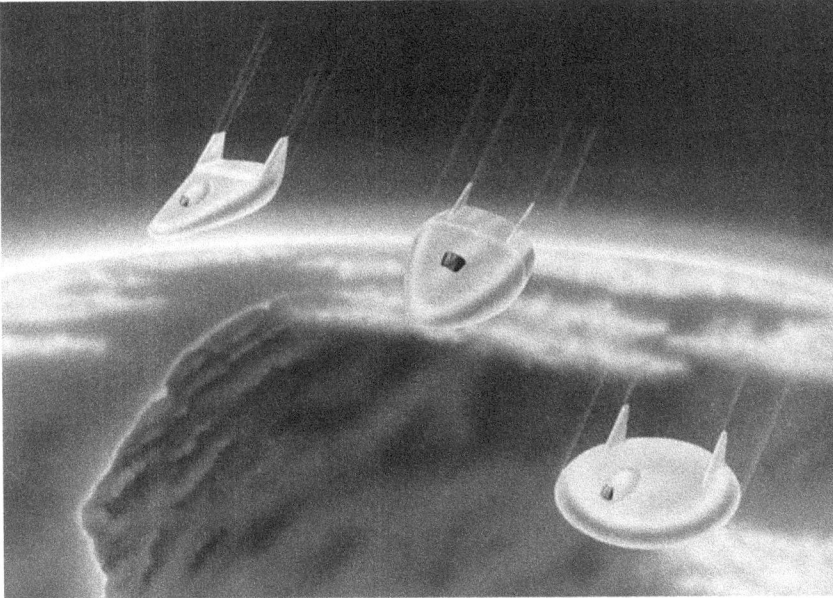

Dale Reed, who inaugurated the lifting body flight research at NASA's Flight Research Center (later, Dryden Flight Research Center, Edwards, CA), originally proposed that three wooden outer shells be built. These would then be attached to the single internal steel structure. The three shapes were (from viewer's left to right) the M2-F1, the M1-L, and a lenticular shape. Milton O. Thompson, who supported Reed's advocacy for a lifting body research project, recommended that only the M2-F1 shell be built, believing that the M1-L shape was "too radical" while the lenticular one was "too exotic." Although the lenticular shape was often likened to that of a flying saucer, Reed's wife, Donna, called it the "powder puff." NASA EC62-175.

this was reflected in their flight profiles. The ASVs were to evaluate materials and structural concepts and to measure temperature, heat flux, and pressure distribution during hypersonic gliding reentry. The ASVs were boosted to altitudes of 190,000 to 225,000 feet and velocities from 16,000 to 19,500 ft/sec with ranges varying from 1,000 to 2,300 miles. On the other hand, the AEVs were boosted to 168,000 and 187,000 feet and a velocity of 13,000 ft/sec, obtaining ranges from 620 to 830 miles. The primary AEV experiments were a 1-by-2-foot body flap on the rear body surface to evaluate unsteady aerodynamic effects on control surfaces and a corrugated columbium panel located ahead of the flap to investigate aeroelastic panel flutter.[32]

32. Charles J. Cosenza, "ASSET: A Hypersonic Glide Reentry Test Program," paper presented at the 1964 Annual Fall Meeting of the Ceramic-Metal Systems Division of the American Ceramic Society; *ASSET*, pp. 5, 11–12.

The first launch (ASV-1) took place on September 18, 1963, just a few months prior to the cancellation of Dyna-Soar. ASSET provided the first real-world reentry exposure for a new generation of refractory materials (like carbon-graphite, thoria-tungsten, and zirconium) and superalloys such as columbium, the cobalt-based alloy L-605, and TZM molybdenum). Advanced thermal-insulating materials were also tested. Bell, Boeing, Martin, McDonnell, Solar, and Vought all provided panels or nose caps that were used on the gliders. Through these tests, ASSET provided a wealth of data that contributed to the development of more advanced materials (primarily composites and carbon-carbon) that would be available when Space Shuttle development began later in the decade.[33]

The ASSET program furnished a great deal of information on the aerother-modynamics and aeroelastic characteristics of a flat-bottom wing-body shape reentering from near-orbital velocities. From 1966 to 1967, the Air Force tested a lifting reentry vehicle under similar conditions. Unlike the structures and heating research-oriented ASSET vehicles, PRIME explored the problems of maneu-vering reentry, including pronounced cross-range maneuvers. ASSET, PRIME, and PILOT (a low-speed piloted demonstrator) were all part of Program 680A, START (Spacecraft Technology and Advanced Reentry Tests).[34]

During late 1964, the Air Force selected the SV-5D for further develop-ment, creating a subtle variation of the basic SV-5 lifting body for the PRIME reentry vehicle. The research and development on this system consumed over 2,000,000 engineering work hours, including configuration and material stud-ies, wind and shock tunnel work, and 50 low-speed glide flights using a recover-able model launched from a ballute.[35]

The PRIME vehicle (often reported as the X-23A, although this appears to be unofficial) was an 890-pound lifting body constructed primarily of 2014-T6 titanium alloy, with some beryllium, stainless steel, and aluminum used where appropriate. The vehicle was completely covered with a nonreceding, charring ablative heat shield consisting of ESA-3560HF for the larger flat surface areas and a more robust ESA-5500M3 for the leading edges. These

---

33. *ASSET: Pioneer of Lifting Reentry*, pp. 486–495.

34. *Advanced Development Program Development Plan for PRIME: A Project Within START, Program 680A* (April 1966); *The X-Planes*, pp. 163–165; "Project PRIME: Hypersonic Reentry," pp. 542–555, 572.

35. "PRIME: A Project Within START"; *The X-Planes*, p. 163; "Program START Hypervelocity Test Program White Paper," Air Force report (no number) (April 6, 1964); "History of the START Program and the SV-5 Configuration," AFSC publication series 67-23-1 (October 1967); "Project PRIME: Hypersonic Reentry," pp. iv–vi.

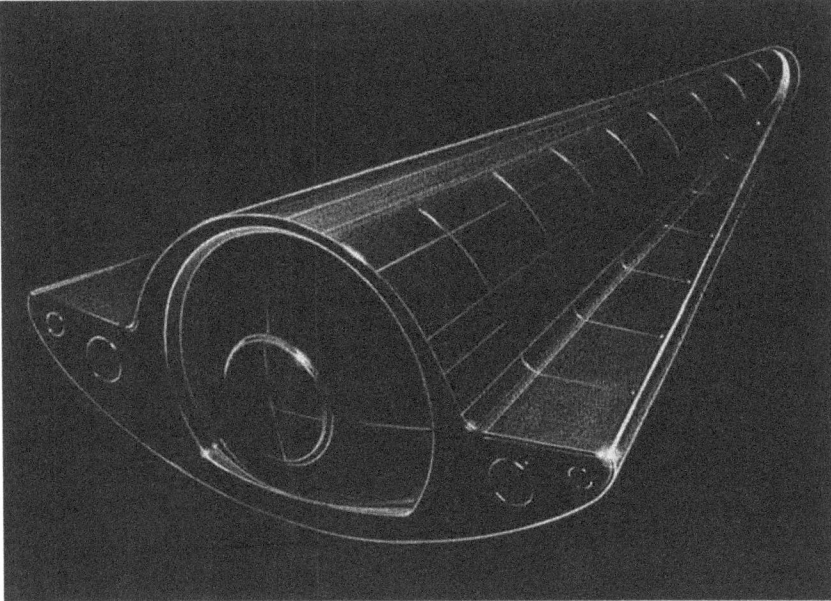

This 1960 concept from General Electric for an Apollo spacecraft bore a striking resemblance to the ASV-1 spaceplane developed for the ASSET program. NASA.

Martin-developed ablators used silica-carbon fibers and siloxane resin inter-laced through a silicone-based honeycomb to hold the char. The thickness of the ablators varied between 0.8 and 2.75 inches, depending on local heating conditions. The nose cap was constructed of carbon-phenolic composite.[36]

The PRIME flights would terminate at approximately Mach 2 with the deployment of a drogue ballute. As the drogue ballute deployed, its cable would slice through the upper structure of the main equipment compartment, where a 47-foot-diameter recovery parachute was stored. Once the recovery chute was deployed, the SV-5D hung in a tail-down attitude awaiting aerial-retrieval by a Lockheed JC-130B Hercules.[37]

On December 21, 1966, the first PRIME vehicle (FV-1) was launched from Vandenberg AFB, CA, on a trajectory simulating a reentry from low-Earth orbit

---

36. "PRIME: A Project Within START"; "Project PRIME: Hypersonic Reentry," pp. 632–657; *The X-Planes*, p. 163; US Patent 4031059, assigned to Martin Marietta Corp. by Eric L. Strauss, June 21, 1977.

37. "PRIME: A Project Within START"; PRIME Flight Test No. 1, Flight Analysis, Martin report ER 14461, March 1967; *The X-Planes*, p. 163; "Project PRIME: Hypersonic Reentry," pp. 694–725; see also Joel W. Powell and Ed Hengevold, "ASSET and PRIME: Gliding Reentry Test Vehicles," *Journal of the British Interplanetary Society* 36 (1983): 369–376.

with a zero cross-range maneuver. The performance of the Atlas launch vehicle and the PRIME spacecraft was nominal, except for a failure of the recovery parachute. The vehicle fell into the Pacific Ocean and was lost. However, all flight objectives other than recovery had been accomplished, and over 90 percent of all possible telemetry had been received.[38] On March 5, 1967, the second vehicle successfully completed a 654-mile cross-range maneuver during reentry, the first for a returning spacecraft. Several stringers failed to be cut during the parachute separation process, and the vehicle was suspended in a manner that the JC-130B could not recover; the second vehicle was lost in the Pacific Ocean, as well.[39]

On April 19, the third SV-5D completed a 710-mile cross-range maneuver and achieved a hypersonic L/D of 1.0 at velocities in excess of Mach 25, only slightly less than the 1.3 predicted. The performance of the PRIME vehicle and all of its subsystems was perfect, and this time everything worked well for the recovery. The waiting JC-130B successfully snagged the SV-5D at 12,000 feet, less than 5 miles from its preselected recovery site. A complete postflight inspection revealed that the vehicle was in satisfactory shape and could be launched again if needed. This was an important, if somewhat unheralded, milestone in demonstrating the potential reusability of lifting reentry spacecraft.[40] Satisfied with the results of the first three flights, the Air Force cancelled the last PRIME flight.

It would be difficult to overestimate the importance of the ASSET and PRIME tests to the development of the Space Shuttle. These nine suborbital tests, along with the hypersonic database generated by the X-15, provided a disproportionate amount of the actual aerothermodynamic data on which the Shuttle orbiter designs were based.

Piloted Lifting Bodies

In the early 1960s Robert Dale Reed, an aerospace engineer at the NASA Flight Research Center, near Edwards AFB, had been following the development of the lifting bodies with considerable interest. Reed noted that while the hypersonic flying qualities of the designs were no longer in question, there was still considerable doubt concerning their low-speed stability. In February

---

38. "PRIME: A Project Within START"; PRIME Flight Test No. 2, Flight Analysis, Martin report ER 14462 (June 1967); *The X-Planes*, p. 163.

39. "PRIME: A Project Within START"; PRIME Flight Test No. 3, Flight Analysis, Martin report ER 14463 (July 1967); *The X-Planes*, p. 163.

40. *The X-Planes*, p. 163; "Project PRIME: Hypersonic Reentry," pp. 700–725.

1962, Reed built a 24-inch model of the M2 shape and launched it from a 60-inch-wingspan, radio-controlled carrier aircraft. During several of the drop tests, his wife, Donna, used an 8-millimeter home movie camera to record the flights, which were later shown to Eggers and the Flight Research Center director, Paul F. Bikle. The results were encouraging enough for Bikle to authorize a 6-month feasibility study of a lightweight piloted M2 glider, construction of which was subsequently authorized in September 1962.[41] The M2 shape was the first shape to be tested, but Reed anticipated testing the M1-L and lenticular shapes as well. This plan eventually was dropped, and the program went straight into the mission weight phase having only tested the M2-F1.[42]

The size of the M2-F1 was dictated by the desire to have a wing loading of 9 pounds per square foot, resulting in a vehicle that was 20 feet long, 10 feet high, and 14 feet wide.[43] Complete with its pilot, the M2-F1 weighed 1,138 pounds.[44] Following ground checkout of the control system, the M2-F1 was trucked to NASA Ames, where it was tested in the 40-by-80-foot, full-scale wind tunnel. On many of the wind tunnel runs, which ultimately totaled about 80 hours, NASA test pilot Milton O. Thompson was seated in the M2-F1, operating the controls while data was collected. The wind tunnel tests at Ames were completed in March 1963, and the M2-F1 was trucked back to the Flight Research Center. The first of many ground tows behind various trucks took place on April 5, 1963. These were intended to lead to captive flights using a Parasev-like canopy, but none of the vehicles proved fast enough to get the M2-F1 airborne.[45]

In response, the Flight Research Center purchased a stripped-down Pontiac convertible with a 455-cubic-inch engine, four-barrel carburetor, and a four-speed stick shift. Ninety-three car tows were performed before the first air tow

---

41. Interview with Dale Reed published in the Dryden Flight Research Center (DFRC) X-Press, March 10, 1967. For Reed's account of the piloted lifting-body program at Edwards, see Reed, with Lester, *Wingless Flight*.

42. *Testing Lifting Bodies at Edwards*; *Developing and Flight Testing the HL-10 Lifting-Body*, p. 6.

43. Terminology dies hard in aerospace, and vehicles without wings still have wing loading, and aircraft without ailerons still do aileron rolls.

44. "Piloted Lifting-Body Demonstrators," p. 869.

45. *Testing Lifting Bodies at Edwards*; V.W. Horton, R.C. Eldridge, and R.E. Klein, "Flight Determined Low-Speed Lift and Drag Characteristics of the Lightweight M2-F1 Lifting-Body," NASA TN-D-3021 (September 1965); John G. McTigue and Milton O. Thompson, "Lifting-Body Research Vehicles in a Low-Speed Flight Test Program," paper presented at the ASSET/Advanced Lifting Reentry Technology Symposium in Miami, FL, December 14–16, 1965, pp. 1–2.

was attempted.[46] Air tows were accomplished behind a Douglas R4D (Navy version of the C-47/DC-3), and the first piloted free flight was on August 16, 1963, with Thompson at the controls. On September 3, 1963, the glider was unveiled to the news media, where it immediately became a hot item in the popular press. The M2-F1 completed over 100 flights and 395 ground tows before being retired on August 18, 1966.[47]

By early 1963, preliminary studies were underway on an air-launched, rocket-powered lifting body built largely with systems and equipment left over from the X-1 and X-15 programs. Engineers intended to expand the envelope of lifting body research into the low-supersonic and transonic speed regions, and to evaluate the landing behavior of a mission-weight lifting body.[48]

In early 1964, Paul Bikle, Dale Reed, and Milt Thompson proposed building two mission-weight versions of the M2 that would be launched by the NB-52 carrier aircraft used by the X-15 program at the Flight Research Center. Officials at NASA Headquarters suggested, however, that the Langley HL-10 shape also be tested. Accordingly, on June 2, 1964, the Flight Research Center awarded a fixed-price contract to the NorAir Division of Northrop.[49]

Interestingly, designers made a conscious decision to size the M2-F2 for a possible future orbital flight-test series. The diameter at the base of the original cone was intentionally made the same diameter as the upper stage of a Titan launch vehicle. It was hoped that by constraining the size of the glider to the mold lines of the booster, the destabilizing influence of a lifting shape could be minimized. This was, at least partly, in response to problems experienced with Dyna-Soar in which wind tunnel tests showed that the glider had an adverse effect on the launch vehicle. Of course, the M2-F2 was not designed to perform an actual reentry, but it was hoped that an identically sized orbital vehicle might be built in the future.[50]

M2-F2

The M2-F2 was rolled out of the Northrop plant on June 15, 1965, and was trucked to Edwards the next day. Of conventional aluminum construction, the

---

46. Reed, with Lester, *Wingless Flight*, pp. 33–35; Hallion, *On The Frontier*, pp. 150–151; *Testing Lifting Bodies at Edwards*.

47. "Piloted Lifting-Body Demonstrators," p. 890; Reed, with Lester, *Wingless Flight*.

48. "Piloted Lifting-Body Demonstrators," pp. 872–873; Reed, with Lester, *Wingless Flight*; Hallion, *On the Frontier*, pp. 150–152.

49. *Testing Lifting Bodies at Edwards*; "Lifting-Body Research Vehicles," p. 2; conversations between Dale Reed and Dennis R. Jenkins, 1998.

50. *Testing Lifting Bodies at Edwards*.

The M2-F1 lifting body under tow at the NASA Flight Research Center. The wingless, lifting body design was initially conceived as a means of landing a spaceplane horizontally after atmospheric reentry. The absence of wings meant that less surface area would need to be protected from reentry heating. In 1962, NASA undertook a program to build a lightweight, unpowered lifting body as a prototype to flight-test the wingless concept. It would look like a "flying bathtub" and was designated the M2-F1, the "M" referring to "manned" and "F" referring to "flight" version. NASA ECN-408.

M2-F2 was 22 feet long, spanned 9.6 feet, and weighed 4,630 pounds without its single Reaction Motors XLR11 rocket engine. A full-span ventral flap controlled pitch, while split dorsal flaps controlled roll (lateral) motion through differential operation and pitch trim through symmetrical operation. Twin ventral flaps provided directional (yaw) control and acted as speed brakes.[51]

On March 23, 1966, the M2-F2 completed its first captive-flight under the wing of the NB-52. However, there was still concern that the lifting body might fly upward and impact the carrier aircraft after it was released, so two wind tunnel test series preceded the first glide flight. A subscale model was tested in the Langley 7-by-10-foot High Speed Tunnel to define the launch transients produced by the flow field of the NB-52 and to establish the proper carry angle for the pylon adapter. These tests predicted that an abrupt, controllable right roll would occur at launch but that the lifting body would fall downward. The second flight was a full-scale test using the M2-F2 in the Ames 40-by-80-foot wind tunnel test section. Data showed that the performance and stability were adequate to begin flight-testing.[52]

---

51. Reed, with Lester, *Wingless Flight: The Lifting-Body Story*, pp. 79–81; "Piloted Lifting-Body Demonstrators," pp. 877–880; *Testing Lifting Bodies at Edwards*.

52. *Testing Lifting Bodies at Edwards*.

Jay L. King, Joseph D. Huxman, and Orion D. Billeter assist NASA research pilot Milton O. Thompson (on the ladder) into the cockpit of the M2-F2 lifting body at the NASA Flight Research Center. The M2-F2 is attached to the same NB-52 wing pylon used by the X-15 program. NASA EC66-1154.

NASA research pilot Thompson made the first unpowered free flight on July 12, 1966, and 12 additional glide flights were made in the following 4 months. On the morning of May 10, 1967, during the 16th glide flight, NASA test pilot Bruce Peterson exited the second of two planned S-turns only to find the M2-F2 rolling and banking wildly. Peterson recovered control but was no longer lined up with the marked runway on the lakebed, and he was too low to make a correction. Peterson successfully completed the flare, and pulled the landing gear deployment handle just as the vehicle touched the lakebed. The M2-F2 rolled and tumbled end over end several times at more than 250 mph before coming to rest upside down on the lakebed.[53] Peterson

---

53. Video footage of this accident was seen weekly at the beginning of the *Six Million Dollar Man* television series. Bruce Peterson became the Director of Safety at the FRC and continued on limited flight status in the Marine Corps Reserves.

was seriously injured and the M2-F2 was essentially destroyed. Fortunately, Peterson made an excellent recovery, although he lost the use of one eye.[54]

Subsequent investigation revealed that settings within the control system allowed Peterson to trigger a severe pilot-induced oscillation during the final approach. Later theoretical analysis would show that this instability was the result of a coupled roll-spiral mode of the pilot-airplane combination.[55] This was the unfortunate end of the M2-F2 flight program, and the wreckage of the M2-F2 was transported to the Northrop facility in Hawthorne, CA, for inspection.

M2-F3

On January 28, 1969, NASA announced the crashed vehicle would be rebuilt as the M2-F3, with a large center vertical stabilizer, in addition to the upswept wingtips. The new stabilizer was not used for additional directional stability, but rather to prevent the strong yawing moment produced when the upper flaps were deflected asymmetrically for roll control.[56]

NASA test pilot William H. Dana took the rebuilt M2-F3 for its first glide flight on June 2, 1970, and found the handling qualities much improved. The M2-F3 made its fastest flight, Mach 1.6, on December 13, 1972, and its highest flight on December 20, when it reached 71,493 feet. The vehicle was retired after completing 43 flights—16 as the M2-F2 and 27 as the M2-F3.[57]

At one point, Northrop submitted an unsolicited $200 million proposal to build an orbital version of the M2 shape launchable by a Titan booster. Interestingly, neither NASA nor the Air Force took action, mainly because there were no funds, and the proposal faded from sight. However, while evaluating this concept, researchers noted that the degradation in performance and stability due to the roughened surface of an ablative heat shield after reentry was not factored into the lifting body flight-test program. Subsequent analysis showed that this

---

54. "Piloted Lifting-Body Demonstrators," pp. 881–883; Reed, with Lester, *Wingless Flight*; Hallion, *On the Frontier*, pp. 158–159.

55. Robert W. Kempel, "Analysis of a Coupled Roll-Spiral Mode, Pilot-Induced Oscillation Experienced With the M2-F2 Lifting-Body," NASA TN D-6496 (September 1971). A coupled roll-spiral mode is a combination of two handling characteristics that are normally quite benign (roll mode and spiral mode). When the two combine into an oscillatory mode, the handling qualities can become unpredictable and potentially uncontrollable.

56. Reed, with Lester, *Wingless Flight*, p. 115; "Piloted Lifting-Body Demonstrators," pp. 882–884.

57. A.G. Sim, "Flight Determined Stability and Control Characteristics of the M2-F3 Lifting Body Vehicle," NASA TN-D-7511 (December 1973); Robert W. Kempel, William H. Dana, and A.G. Sim, "Flight Evaluation of the M2-F3 Lifting Body Handling Qualities at Mach Numbers from 0.30 to 1.61," NASA TN-D-8027 (July 1975).

NASA research pilot John A. Manke is seen here in front of the M2-F3 lifting body in 1972. NASA ECN-3448.

degradation was sufficient to raise serious questions about the true ability to land any of these vehicles following an actual reentry, and limited testing eventually took place using the HL-10 in the Ames full-scale wind tunnel.[58]

HL-10

Concurrently with the M2 flight tests, the Flight Research Center was also flying the HL-10. This vehicle was 22 feet long and spanned almost 15 feet

---

58. *Testing Lifting Bodies at Edwards.*

across its aft fuselage and also was conceived with the intent of developing an orbital spaceplane concept. Internally, the M2-F3 and HL-10 were similar, with nearly identical subsystems and structural details. Unlike the other lifting bodies, the HL-10 did not incorporate a canopy, and the pilot was completely dependent on the nose and side windows for visibility during landing. The control system consisted of upper-body and outer stabilizer flaps for transonic and supersonic trim, blunt trailing-edge elevons, and a split rudder on the central vertical stabilizer. Since this was a test vehicle not intended for space flight, the thermal protection system was minimal.[59]

As with the M2-F2, subscale tests were conducted in the Langley 7-by-10-foot High Speed Tunnel to define the launch transients produced by the flow field of the NB-52 and the proper carry angle for the pylon adapter. The HL-10 was rolled out in Hawthorne, CA, on January 18, 1966, and was subsequently trucked to the Ames 40-by-80-foot wind tunnel test section for final testing. The results were generally satisfactory, although some momentary flow separation was noted on the tip stabilizers under certain flight conditions. Engineers at Ames and Northrop did not consider this particularly serious and cleared the HL-10 for flight.[60]

Peterson took the HL-10 on its first glide flight on December 22, 1966, revealing that the flow separation was much more serious than initially thought, and NASA Langley embarked on yet another round of wind tunnel tests. The solution was to add an inward-cambered glove to the leading edge of the tip stabilizers that allowed the airflow to stay attached at high angles of attack and low speeds. The HL-10 was grounded for 15 months while the problem was identified and corrected.[61]

During its next flight, on March 15, 1968, research pilot Jerauld R. Gentry found the modifications worked well and the HL-10 handled nicely. On November 13, research pilot John A. Manke made the first successful powered flight, and on May 9, 1969, the HL-10 made the first supersonic flight of any of the piloted lifting bodies. Air Force test pilot Peter C. Hoag reached Mach 1.86 on February 18, 1970, and 9 days later Dana topped out at 90,303 feet,

59. *Northrop: An Aeronautical History*, pp. 234–235; "Piloted Lifting-Body Demonstrators," pp. 885–886; Hallion, *On the Frontier*, pp. 160–167; *Testing Lifting Bodies at Edwards*.

60. "Flight Test Results Pertaining to the Space Shuttlecraft"; "Piloted Lifting-Body Demonstrators," pp. 885–886; *Testing Lifting Bodies at Edwards*.

61. M.H. Tang and G.P.E. Pearson, "Flight-Measured HL-10 Lifting Body Center Fin Loads and Control Surface Hinge Moments and Correlation with Wind Tunnel Predictions," NASA TM-X-2419 (October 1971); Reed, with Lester, *Wingless Flight*, p. 102–123; "Piloted Lifting-Body Demonstrators," pp. 885–886; *Developing and Flight Testing the HL-10 Lifting-Body*.

HL-10 lifting body model in NASA Langley Research Center's Full-Scale Tunnel in 1963. NASA L-1963-06125.

making the HL-10 the fastest and highest flying of the lifting bodies. It was on this flight that the HL-10 for the only time approached a situation in which the thermal protection system came into play in the slightest, but the reality was that this concern could wait for the development of future systems. The basic research program consisted of 35 flights, and the HL-10 was deemed the best performer of the lifting bodies. The shape later became the basis for several early Space Shuttle concepts.[62]

Two additional HL-10 flights were undertaken to determine whether turbo-jet engines were needed to land a spaceplane returning from space, or if it could be accomplished with a glider approach. Three Bell Aerosystems 500-pound force hydrogen peroxide engines replaced the XLR11 rocket. These engines fired as the vehicle passed through 6,500 feet on the way to landing, reducing

---

62. "Flight Test Results Pertaining to the Space Shuttlecraft"; "Piloted Lifting-Body Demonstrators," p. 886; Reed, with Lester, *Wingless Flight;* Hallion, *On the Frontier*, pp. 161–163.

COMPARISON OF REFRACTORY AND ABLATIVE SHINGLE STRUCTURES
HL–10 SPACECRAFT

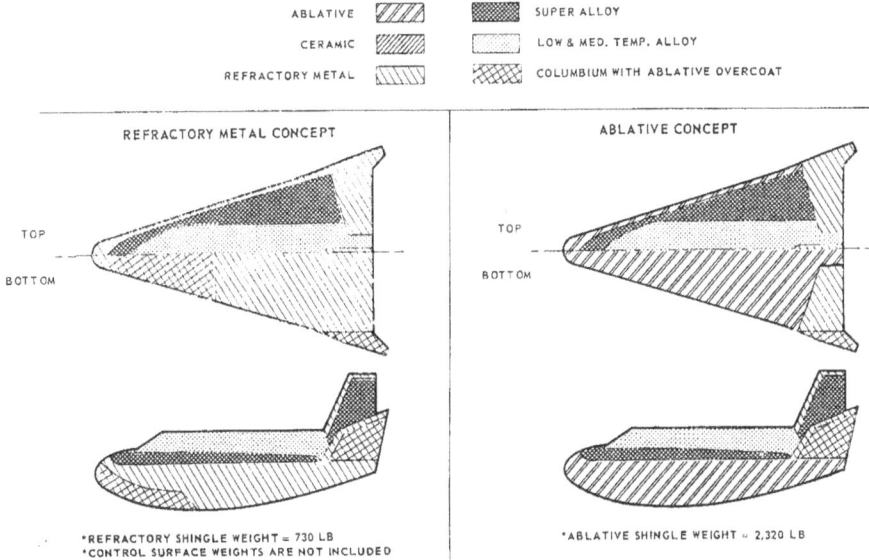

ABLATIVE		SUPER ALLOY
CERAMIC		LOW & MED. TEMP. ALLOY
REFRACTORY METAL		COLUMBIUM WITH ABLATIVE OVERCOAT

REFRACTORY METAL CONCEPT

TOP

BOTTOM

*REFRACTORY SHINGLE WEIGHT = 730 LB
*CONTROL SURFACE WEIGHTS ARE NOT INCLUDED

ABLATIVE CONCEPT

TOP

BOTTOM

*ABLATIVE SHINGLE WEIGHT = 2,320 LB

Comparison of refractory and ablative shingle structures for a proposed orbital HL-10 thermal protection system. NASA.

the approach angle from the normal 18 degrees to 6 degrees and boosting the airspeed to over 350 mph. At 200 feet above the lakebed, the pilot would shut down the engines, extend the landing gear, and make a routine landing. Hoag made flights on June 11 and July 17, 1970, marking the last flight of the HL-10 in testing this approach to landing.[63]

Hoag's comments about the landing engines were quite negative. He reported that the shallow approach made it difficult for him to judge where the vehicle would touchdown. He also reported that the nose-high attitude during the entire maneuver forced the pilot to rely on the nose window, which provided poor depth perception, for a long time before touchdown. The consensus was that the weight and complexity of the engine installation and the increased pilot workload outweighed the benefits of landing engines—namely, the ability to make a missed approach, due to shallower descent angles and

---

63. L.W. Strutz, "Flight-Determined Derivatives and Dynamic Characteristics for the HL-10 Lifting Body Vehicle at Subsonic and Transonic Mach Numbers," NASA TN-D-6934 (September 1972); "Piloted Lifting-Body Demonstrators," p. 887; Reed, with Lester, *Wingless Flight*.

higher approach speeds.[64] It was an important lesson for the future of space-planes, and accordingly NASA proceeded with plans for the Space Shuttle to perform a glider landing on a runway.

Air Force Efforts

Although heavily involved in supporting the M2-F2 and HL-10 programs, the Air Force also conducted independent research into lifting body concepts for reentry and recovery from space. In fact, as early as the mid-1950s, the AFFDL had begun investigating lifting body designs (SAMOS, SAINT, etc.) for a variety of purposes, including for use as a film-return vehicle for reconnaissance satellites. The laboratory conducted a detailed evaluation of a variety of shapes beginning in 1959 and continuing until the creation of X-24B. Most of these shapes tended to reflect the relatively high L/D preferred by researcher Draper.[65]

As early as 1962, Draper proposed flight-testing piloted prototypes of several of the designs, but the Air Force, much to the dismay of both Draper and NASA, declined. Nevertheless, Draper and his associates proceeded with their laboratory studies, generating a large series of tailored shapes. Some of these reentered the atmosphere as lifting bodies but deployed variable-geometry wings to increase their transonic L/D. The laboratory also explored interference configurations that used complex undersurface designs to position shock flows for favorable increases in lift.[66] Similarly, the laboratory investigated Nonweiler caret-wing wave rider configurations that made use of favorable flow interference to increase their hypersonic L/D.[67] This approach continued to be pursued during 1991 in support of the X-30 NASP, prior to its cancellation.[68]

---

64. Weneth D. Painter and G.J. Sitterle, "HL-10 Lifting Body Flight Control System Characteristics and Operational Experience," NASA TM-X-2956 (January 1974).

65. *Evolution and Development of Hypersonic Configurations*, pp. 8–9.

66. The North American XB-70A bomber used a somewhat similar technique, folding its wingtips down at high speeds to capture the shock wave generated by its nose. This increased the overall L/D and improved directional stability. Although successful on the XB-70A, it would prove a great deal more difficult to use this technique as speeds increased into the hypersonic regime because of thermal considerations.

67. Dr. Terence R.F. Nonweiler at the Queen's University in Belfast wrote the first paper on the caret-wing wave rider concept in 1951.

68. "Piloted Lifting-Body Demonstrators," pp. 894–900.

X-24A

While NASA was testing the M2 and HL-10, the Air Force was building a piloted lifting body demonstrator. Project PILOT (piloted low-speed tests) was part of the ongoing START program, which also included the ASSET and PRIME vehicles. This would provide the last part of a knowledge base that would capture data for the A3-4/SV-5 shape from the Mach 25 achieved by the SV-5D (X-23A) all the way to approach and landing with the SV-5P (X-24A). The objectives of PILOT were to investigate static and dynamic stability characteristics and to verify control techniques for low-supersonic, transonic, and landing-speed regimes.[69]

Martin Marietta was awarded a contract on March 2, 1966, for a single-rocket-powered SV-5P, designated the X-24A. Construction of the vehicle consumed a little over 12 months under the direction of Bastian "Buzz" Hello and Lyman Josephs.[70] When the X-24A was rolled out on August 3, 1967, it was 24 feet long, spanned 13 feet, and was essentially four times larger than the PRIME SV-5D shape, with a revised canopy area. Power was supplied by the seemingly irreplaceable XLR11. The vehicle had an empty weight of less than 6,000 pounds, increasing to 11,000 pounds with propellants and a pilot. Since testing was to be limited to less than Mach 2, the X-24A was of conventional aluminum construction, with no special attention paid to heat protection.

The basic aerodynamic control system consisted of eight movable surfaces. Pitch control was derived from the symmetrical deflection of the lower and/or upper flaps, depending upon flight conditions. Differential deflection of the flaps provided the primary roll control. Pitch or roll commands, which caused either lower flap to close fully, resulted in the transfer of control to the corresponding upper flap through a simple clapper mechanism. The two pairs of rudders were deflected symmetrically as a bias feature with directional control provided by the deflection of the surfaces in unison.[71]

The X-24A was added to the existing cadre of lifting bodies (the M2-F1, M2-F2, HL-10) in the NASA Dryden Flight Research Center hangar, where they were being flight-tested by a joint Air Force/NASA test team. The X-24A arrived at Edwards on August 24, 1967, for initial ground testing, but the vehicle was soon airlifted to NASA Ames for testing in the 40-by-80-foot wind tunnel test section. While at Ames, additional tests were performed with a

---

69. *The X-Planes*, p. 165; "Piloted Lifting-Body Demonstrators," pp. 901–903.

70. *Testing Lifting Bodies at Edwards*; "Piloted Lifting-Body Demonstrators," p. 918. Hello would later lead the team at Rockwell International that bid on, and won, the Space Shuttle orbiter contract.

71. "Piloted Lifting-Body Demonstrators," p. 919; Hallion, *On the Frontier*, pp. 163–165; *Testing Lifting Bodies at Edwards*.

This 1968 photo shows the bulbous X-24A lifting body on the lakebed adjacent to the NASA Flight Research Center. The X-24 was one of a group of lifting bodies flown by the NASA Flight Research Center in a joint program with the U.S. Air Force at Edwards AFB from 1963 to 1975. NASA ECN-2006.

simulated rough ablator applied to the vehicle. Water-soluble glue was sprayed on the vehicle, and a wire-mesh screen was laid over the surface to simulate the pattern of the ablator honeycomb. Coarse sand was then sprayed over the surface. When the glue was dry, the wire mesh was removed, leaving a rough protruding pattern on the surface. It was hoped that these results could be correlated with the effects that were observed on the PRIME vehicle following an actual reentry. The results from the full-scale tunnel tests showed a 20-percent reduction in maximum L/D and some degradation in stability caused by surface roughness.[72] As a result, any thoughts of flying and landing the vehicle with the simulated ablator roughness were quickly dismissed. Although this was an interesting data point, and used during the development of the Space

---

72. Jon S. Pyle and Lawrence C. Montoya, "Effects of Roughness of Simulated Ablated Material on Low-Speed Performance Characteristics of a Lifting-Body Vehicle," NASA TM-S-1810 (1969); *Testing Lifting Bodies at Edwards.*

Shuttle, no further work on it was done using the X-24A.[73] However, it did raise doubts about the usefulness of an ablative heat shield on any operational lifting reentry vehicle that would be landed horizontally, and it eventually pointed R&D efforts toward the thermal protection system used on the Space Shuttle with its ceramic tiles and reinforced carbon-carbon.[74]

The X-24A arrived back at Edwards on March 15, 1968, but it was not until a year later that the vehicle was cleared for flight. Finally, on April 17, 1969, Air Force test pilot Gentry made the first of nine glide flights. Gentry flew the first powered flight on March 19, 1970, and during the course of its test program, the X-24A made 28 flights, reaching a maximum speed of Mach 1.60 and an altitude of 71,400 feet. The last portion of the X-24A program was dedicated to simulating Space Shuttle approaches and landings, and the vehicle completed its last flight on June 4, 1971.[75]

X-24B

While the X-24A was still being built, Martin also completed two jet-powered low-speed lifting bodies based on the SV-5 shape as a company-funded venture for use as astronaut trainers at the Aerospace Test Pilot School. However, the Air Force considered the vehicles significantly underpowered and they never flew. During 1968, the Air Force Systems Command solicited suggestions for possible uses for the two unused SV-5J airframes, and the AFFDL responded on January 23, 1969, with a proposal by Draper and Bill Zima to modify one of them into the FDL-7 configuration. To minimize modification costs, it was proposed that the basic SV-5J body structure and vertical stabilizers be retained, with the FDL-7 shape being gloved over the existing forebody, creating the FDL-8. A flat-bottom, 72-degree sweep, double-delta configuration contributed to a high hypersonic L/D, while a 3-degree nose ramp provided the proper hypersonic trim condition. Despite tailoring the airframe for hypersonic velocities, the X-24B was never intended to fly that fast.[76]

The laboratory saw this as an opportunity to perform a low-cost flight demonstration on one of their configurations with a hypersonic L/D of 2.5, in contrast to the other lifting bodies that had hypersonic L/Ds of between

---

73. "Piloted Lifting-Body Demonstrators," p. 919; Hallion, *On the Frontier*, pp. 163–165; *Testing Lifting Bodies at Edwards.*

74. "Effects of Roughness of Simulated Ablated Material"; "Piloted Lifting-Body Demonstrators," p. 919; Hallion, *On the Frontier*, pp. 163–165.

75. *The X-Planes*, pp. 166–167; "Piloted Lifting-Body Demonstrators," pp. 919–922.

76. *Evolution and Development of Hypersonic Configurations*, p. 50; "History of the Air Force Plant Representative; Piloted Lifting-Body Demonstrators," p. 923; *Testing Lifting Bodies at Edwards.*

1.2 and 1.4. The proposed jet-powered FDL-8 was to be air-launched from one of the NB-52 carrier aircraft, but as the studies matured, the advantages of rocket propulsion became apparent, and Draper soon proposed using the X-24A instead of the SV-5J to reduce the costs of the program.[77] When the X-24B (also known as the SV-5P-2) was rolled out on October 11, 1972, the vehicle had grown over 14 feet in length and 10 feet in span, its weight was up to 13,700 pounds, and the center of gravity was considerably farther forward relative to the main landing gear.[78]

The X-24B was delivered to Edwards on October 24, 1972, for several months of ground testing. NASA test pilot Manke made the first glide flight on August 1, 1973, and the first powered flight on November 15. On October 25, 1974, Air Force test pilot Michael V. Love made the fastest X-24B flight at Mach 1.76, and Manke took the vehicle to 74,130 feet on May 22, 1975, marking its highest flight.[79]

By the summer of 1975, the Space Shuttle designers were again debating whether to provide air-breathing landing engines. The primary concern was whether low-L/D reentry shapes could successfully complete unpowered landings on conventional runways. From the beginning, engineers believed that the X-24B could accomplish this, and nine accuracy-landing flights showed that pilots could successfully touch down within 500 feet of a designated marker on the lakebed runway. Since the vehicle was traveling at about 220 mph at landing, the pilot was within the touchdown zone for only 3 seconds—a remarkable demonstration of energy management.[80]

Based on these results, research pilots Love and Manke proposed landing on the 15,000-foot concrete runway at Edwards, something that had not been done with the X-15 or earlier lifting bodies due primarily to their lack of ground-steering capability. Johnny Armstrong, the X-24B program manager, presented the plan to the commander of the Air Force Flight Test Center, Major General Robert A. Rushworth. The general was well acquainted with

---

77. *Evolution and Development of Hypersonic Configurations*, p. 54; "Piloted Lifting-Body Demonstrators," p. 923; *Testing Lifting Bodies at Edwards*.

78. *Testing Lifting Bodies at Edwards*; "Piloted Lifting-Body Demonstrators," pp. 923–924; Hallion, *On the Frontier*, pp. 165–167.

79. "Piloted Lifting-Body Demonstrators," pp. 925–928; Hallion, *On the Frontier*, pp. 165–167; Jerome C. Brandt, "XLR11-RM-13 Rocket Engine Development and Qualification Program," AFFTC report FTC-TD-69-1 (August 1969); Jerome C. Brandt, "X-24 Propellant System Development and Qualification Program," AFFTC report FTC-TD-69-7 (August 1969).

80. John L. Stuart, "Analysis of the Approach, Flare, and Landing Characteristics of the X-24B Research Aircraft," Air Force report AFFTC-TR-76-9 (November 1977).

The X-24B flying over the lakebed at NASA Dryden Flight Research Center in 1975. The X-24B was the last of the lifting bodies to fly, after being modified from the original X-24A. The X-24B's design evolved from a family of potential reentry shapes, each with higher lift-to-drag ratios, proposed by the Air Force Flight Dynamics Laboratory. NASA EC75-4642.

unpowered landings, having flown the X-15 more than any other pilot (34 times). Armstrong remembers Rushworth was supportive but cautioned, "I will forgive you if you land long and end up rolling onto the lakebed, but landing short of the concrete runway will be unacceptable."[81]

The desired touchdown point was a stripe painted on Runway 04, approximately 5,000 feet from the approach end. Manke performed the first runway landing on August 5, 1975, and landed in a slight bank with one main wheel touching before the stripe, and the other wheel touching after the stripe. Love made the second runway landing on August 20, 1975, touching down 400 feet beyond the stripe. Both pilots reported that the additional visual cues provided by roads and Joshua trees made the final phase of the landing easier than on the smooth, flat lakebed. The two runway-landing rollouts were about 34 percent longer than equivalent lakebed landings, and a similar difference

---

81. Conversations by the author with Johnny Armstrong and Bob Hoey, related on Armstrong's Web site, *http://members.aol.com/afftc/X-24blanding.htm*, accessed November 11, 2005.

was later noted between lakebed and runway landings of the Space Shuttle during flight tests.[82]

On September 23, 1975, Dana, who had been the last pilot to fly the X-15, also completed the last powered X-24B flight, marking the end of rocket-powered research flights at Edwards AFB.[83] Following this flight, the Air Force decided to provide six checkout flights for three test pilots who had not had the opportunity to fly any of the lifting bodies. These flights included an unrelated test where access panels on the lower surface were used to evaluate Space Shuttle thermal protection system tiles. One array of tiles was bonded to the panel to assess the ability of the tiles to withstand aerodynamic shear forces in flight and to investigate the susceptibility to damage during landings on the dry lakebed. The last flight was on November 25, 1975.[84] In all, the X-24B flight program had consisted of 24 powered and 12 glide flights.

Given the state of the art in 1970, the lifting bodies were not necessarily suitable as spacecraft. While the lifting body flight-test data was being gathered, the effects of ablation surface roughness on low-speed drag were also being evaluated, and full-scale wind tunnel tests of the X-24A with a simulated rough ablator surface showed a 20-percent reduction in L/D. Wind tunnel tests at Wright-Patterson AFB on an 8-percent model of the X-24A showed similar results. Flight tests of the X-15A-2, which used a thin ablative coating, showed a reduction in L/D of about 15 percent after a relatively mild exposure to the aerodynamic heating environment. Comparison tests of two PRIME vehicles, one before flight and one after flight, showed a 30-percent reduction in L/D. These effects were also accompanied by reductions in stability that could obviously be quite detrimental to the handling qualities.[85]

---

82. *Testing Lifting Bodies at Edwards*; Robert G. Hoey et al., "Flight Test Results From the Reentry and Landing of the Space Shuttle Orbiter for the First Twelve Orbital Flights," Air Force report AFFTC-TR-85-11 (June 1985).

83. At the end of the 1990s, it appeared that rocket-powered research vehicles would again fly above Edwards, but the demise of the X-33 and X-34 programs proved otherwise. The X-43A scramjet vehicles, although based at Edwards, were flown over the Pacific Ocean off the coast of California.

84. *Testing Lifting Bodies at Edwards.*

85. Jon S. Pyle and L.C. Montoya, "Effect of Roughness of Simulated Ablated Material on Low Speed Performance Characteristics of a Lifting-Body Vehicle," NASA TM S-1810 (July 1969); Lawrence G. Ash, "Flight Test and Wind Tunnel Performance Characteristics of the X-24A Lifting-Body," vol. II, Air Force report FTC-TD-71-8 (June 1972); John S. Spisak, "Ablation Effects on the Aerodynamics of a Flight Tested Lifting Reentry Vehicle," Air Force report AFFDL FXS TM 71-3 (December 1971).

It must be concluded that the first three lifting body vehicles, as originally conceived, would probably not have been landable following a reentry with a normally ablated thermal protection system.[86] Nevertheless, the lightweight ceramic-tile technology developed for the Space Shuttle opened the door to all of these lifting reentry concepts, and highly maneuverable entries with over 2,400 miles cross range are theoretically possible with X-24B-like configurations. The lifting bodies, however, presented other challenges for designers, such as how to package propellant tanks and other subsystems into the oddly shaped fuselages.

Orbital Lifting Bodies

During the early and mid-1960s, the lifting body seemed like an ideal platform for an orbital vehicle, and, not surprisingly, several proposals were floated to develop such a vehicle. One of these was a study conducted by Martin Marietta for the Langley Research Center using the basic HL-10 shape. This study, led by R.H. Lea, was completed in February 1966 and did not so much concentrate on the changes to a lifting body required for orbital flight but, rather, was a sales tool for the Titan III family of launch vehicles then under development.[87]

Nevertheless, the results of this orbital lifting body study looked promising enough that NASA Langley issued a contract to Martin in April 1966 to study the concept further. Robert L. Lohman was the study manager, and the final report was delivered in May 1967. Parametric design and performance data was developed for five different lifting body sizes (1, 2, 4, 6, and 8 men) and four different launch vehicles (the Titan GLV, a 2- and 5-segment Titan IIIC, and a Saturn IB).[88]

Ultimately, Martin believed that a 25-foot-long, 3-person lifting body weighing 12,342 pounds would be the most cost-effective research tool. The proposed vehicle included a conventional 2219T-6 aluminum airframe and

---

86. Further study and wind tunnel testing were required to identify the true cause of these effects. It is likely that the judicious use of smooth, high-temperature materials (such as carbon-carbon) placed in critical locations on the vehicle would have substantially improved the low-speed characteristics after reentry.

87. R.H. Lea, "Titan II Boosters for Langley's Manned Lifting-Body Reentry Program," Martin-Denver report M-66-8 (February 1966), p. 1.

88. Robert L. Lohman et al., "Study of the Influence of Size of a Manned Lifting-Body Reentry Vehicle on Research Potential and Cost," vol. III, Martin-Baltimore report CR-66354 (May 1967), pp. iii–iv. The contract was (NAS1-6209).

a refurbishable ablative heat shield designed for all-turbulent heating over its entire surface. The vehicle could carry a crew of six by removing some of the research instrumentation. As part of the study, Martin completed a detailed preliminary design of the vehicle and its subsystems, and a minor amount of effort was expended on possible modifications to support rendezvous, docking, and superorbital (34,000 ft/sec) entry. Although the study was based on the HL-10 shape, it was felt that many of the conclusions were equally applicable to the M2 and SV-5 vehicles.[89]

Two nonpiloted flights would be followed by a series of nine piloted flights at a cost of $1 billion. The flights would be launched into a 92-by-230-mile orbit from Cape Canaveral and recovered at Edwards AFB, with Eglin AFB, FL, used as a contingency landing site. Martin believed the first nonpiloted suborbital flight could occur 35 months after the authority to proceed. Four piloted vehicles would be manufactured, and each could be refurbished and reflown every 129 days. It was an aggressive program, and as happened so often in high-technology development programs, the engineers were too optimistic in their enthusiasm for the system and its capabilities.[90]

In the end, there would be no serious attempt to launch a lifting body into orbit, although researchers at both Langley and the Flight Research Center thought such an exercise would be useful. However, neither the Air Force nor NASA had sufficient funds to do so given the large expenditures on Apollo and later on the Space Shuttle program.

BoMi and Brass Bell

Major General (Dr.) Walter R. Dornberger had been the military director of the German Peenemünde test site during World War II, and Dr. Krafft A. Ehricke was a close associate. Both men came to the United States in 1947 and worked for the Government on various missile-related programs. In 1950, Dornberger joined the staff of Bell Aircraft, and Ehricke followed in 1952. While at Bell, they advocated the development of a Silverbird-like skip-glide vehicle. Early Bell skip-glide designs were simple variations of the Sänger-Bredt Silverbird, but the triangular-section straight wings soon began to morph into the delta planform more typical of later concepts. During these studies, the company devoted a great deal of attention to thermal protection systems and investigated both passive and active techniques in considerable detail.

---

89. "Study of the Influence of Size," vol. III, pp. iv, 1, 53, and vol. VII, p. 3.
90. "Study of the Influence of Size," vol. V, pp. vii–viii, 3–14.

On April 17, 1952, Bell proposed a 1-year study of a piloted Bomber-Missile (BoMi) consisting of a five-engine rocket booster with a three-engine, double-delta skip-glider. The two-person booster was nearly 120 feet long with a 60-foot wingspan and was to be manufactured of aluminum alloys with a titanium hot-structure wing leading edge. An active cooling system kept the airframe within its thermal limits by circulating water between a thin outer alloy skin and a heavier inner aluminum skin. A heat exchanger cooled the water and jettisoned superheated vapor overboard. The single-seat skip-glider was 60 feet long with a 35-foot wingspan and carried a 4,000-pound nuclear weapon in a rear-ejecting bomb bay. There appeared to be no military useful-ness for an orbital vehicle, so this suborbital design was capable of 2,650 mph at 100,000 feet with a 3,300-mile range. The pair took off horizontally, and the booster provided power for 2 minutes before the skip-glider separated and ignited its engines. The booster flew back to the launch site for a runway landing and subsequent reuse. The glider then skipped across the atmosphere following the path described by Sänger and Irene Bredt for the Silverbird.[91]

An initial Air Force review of the suborbital BoMi on April 10, 1953, uncovered several deficiencies. It was difficult to envision how the vehicle could be adequately cooled, despite a proposed water-wall concept, and there was insufficient information concerning stability, control, and aeroelasticity at the proposed speeds. The projected range of 3,300 miles was too short for intercontinental operations, minimizing its usefulness as either a bomber or a reconnaissance vehicle.[92]

Notwithstanding these misgivings, on April 1, 1954, the Air Force awarded Bell a 1-year study contract as project MX-2276. Despite the lack of intercon-tinental range and the difficult technological questions, the skip-glide concept seemed to offer significant strategic potential and, at a minimum, it would pro-vide a useful test vehicle for future hypersonic programs. However, the Air Force requirements for MX-2276 were a significant increment beyond what Bell had proposed. The vehicle was to be capable of bombardment and reconnaissance missions with a maximum velocity of 15,000 mph at 259,000 feet and a range of 12,000 miles. The MX-2276 contract was not, per se, a hardware-development

---

91. Clarence J. Geiger, "Strangled Infant: The Boeing X-20 Dyna-Soar," *Case II of The Hypersonic Revolution: Case Studies in the History of Hypersonic Technology, Volume 1, From Max Valier to Project PRIME (1924–1967)* (Bolling AFB, DC: USAF Histories and Museums Program, 1998), pp. 188–189.

92. Headquarters Wright Air Development Center (WADC) to Headquarters Air Research and Development Command (ARDC), "Rocket Bomber Feasibility," April 10, 1953, copy in possession of authors.

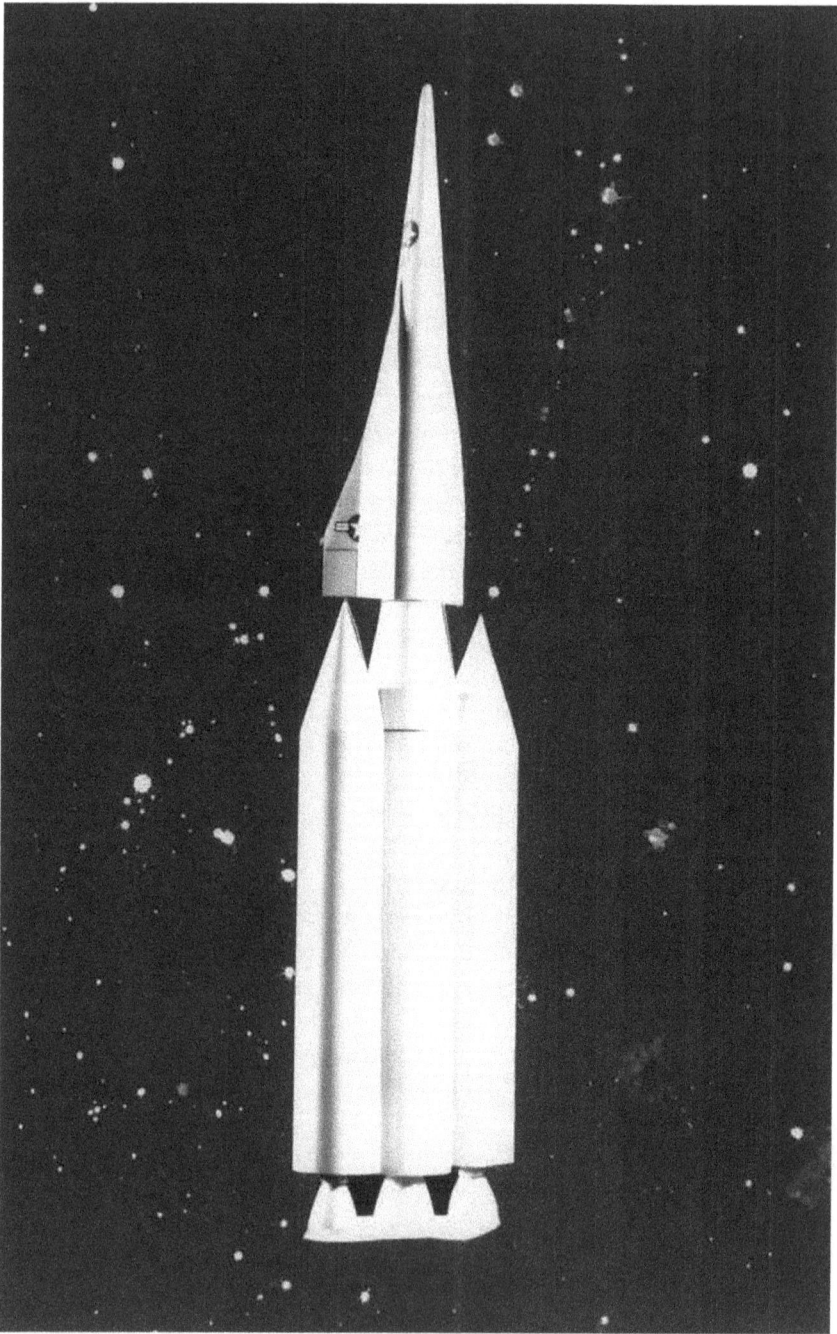

The BoMi (Bomber Missile) was a "boost-glide" concept not unlike the Eugen Sänger's earlier efforts, officially designated the MX-2276. USAF.

effort. Instead, Bell was to define requirements for future programs, with a focus on such issues as the necessity for a piloted space vehicle, possible mission profiles, structural performance at high temperatures, and the feasibility of various guidance systems.[93]

During early 1955, the Air Force asked the NACA to evaluate the BoMi concept. Largely in response to the Air Force request, Allen and Alvin Seiff at NACA Ames conducted a classified comparison of the skip-glide and boost-glide techniques, building on Allen's 1954 evaluation. The conclusion was that the skip-glider had a slightly longer range for any given initial velocity than an equivalent boost-glide concept. However, the researchers again determined that the skip-glider would encounter significantly higher heating rates and experience greater magnitude and longer duration g-loads during reentry. This was because the transfer of kinetic energy to heat occurred in abrupt pulses during the skipping phases of flight—the first skip, of course, was the most severe, but all of the early skips had to endure extreme heating. The high heating rate necessitated more thermal protection, creating a heavier vehicle, and probably negated the small advantage in predicted range, most likely presenting unsolvable technical challenges. The boost-glider, on the other hand, gradually converted kinetic energy to heat over the entire entry trajectory, resulting in a relatively low, constant heating rate, but it still created a substantial total heat load. The August 1955 report concluded that the boost-glide concept "appeared to hold the most promise for an operational military system but noted that it would be of questionable usefulness since it could not…be a very maneuverable vehicle in the usual sense and thus is, perhaps, limited to use for bombing and reconnaissance." Those, fortunately, were the missions that interested the Air Force.[94]

Based largely on the Ames report, on September 30, 1955, Dr. Ira H. Abbott, the NACA Assistant Director for Research, reported to the Air Force that the NACA was generally supportive of the boost-glide concept but believed that more research was required before a development program could be initiated. Abbott reported that the NACA was less enthralled with the skip-glide concept, believing it to be essentially unworkable due to concerns about reentry and recovery from space, especially regarding atmospheric heating, structural stability at high temperatures, and guidance and control of a vehicle at extreme speeds. With encouragement (and, likely, funding) from the Air Force, the

---

93. R&D Project Card, System 464L, August 23, 1957, p. 9. Bell used the BoMi and MX-2276 designations seemingly interchangeably; it appears they were usually the same vehicle at most points during the studies.

94. Seiff and Allen, "Some Aspects of the Design of Hypersonic Boost-Glide Aircraft," pp. 2–21.

NACA undertook further investigation into the boost-glide concept, with Allen and Neice issuing another report in March 1956.[95]

The Ames reports marked the end for the dynamic-soaring technique first proposed by Sänger, and almost all work on skip-gliding ceased in favor of boost-glide concepts. Bell soon replaced the BoMi skip-glider with a generally similar boost-glide concept.

It should be noted that although most early lifting reentry vehicles were called gliders, this description was only appropriate because they were not continually powered. With some exceptions, all of these vehicles carried turbojet engines that would extend the terminal portion (landing) of their flight and frequently included rocket engines to act as upper stages to achieve higher orbits or to help the vehicle maneuver in space.

Rocket-Bomber

Near the same time, at the request of Trevor Gardner, the Assistant Secretary of the Air Force for Research and Development, several boost-glide presentations were given to Air Force Headquarters in November 1955. The concept met with a warm reception since it offered a chance to leapfrog the technology bottleneck that was hampering development of high-speed, long-range bombers.[96]

In response, on December 19, 1955, the Air Force asked the aerospace industry to investigate the concept of a hypersonic bomber. Boeing, Convair, Douglas, North American, and Republic ultimately responded, and three companies—Convair, Douglas, and North American—were awarded 1-year study contracts. All five companies—later joined by Bell, Lockheed, and Martin— also used company funds for the study.[97] However, midway through the study, on June 12, 1956, the Air Force released SR-126 to determine the feasibility of an intercontinental bombardment and reconnaissance system that could circumnavigate the globe at 100,000 feet altitude, with a maximum altitude of 260,000 feet. This piloted hypersonic Rocket Bomber (RoBo) needed to be operational by 1965 and superseded the earlier study.[98]

---

95. Dr. Ira R. Abbott, NACA Assistant Director for Research, to Headquarters WADC, "Requested Comments on Project MX-2276," September 30, 1955, copy in possession of authors.

96. "Directorate of Systems Management Weekly Activity Report," November 25, 1955, and December 15, 1955.

97. R&D Project Card, System 464L, August 23, 1957, p. 9.

98. "SR-126, Manned Hypersonic Rocket Bomber Study" (June 12, 1956); R&D Project Card, System 464L, August 23, 1957, p. 2.

In late July, 1957, Dornberger presented "An Approach to Manned Orbital Flight" to the Air Force Scientific Advisory Board; the information that was presented relied heavily on the boost-glide designs Bell had been investigating for the previous 5 years. During the presentation, Dornberger stated that engineering solutions have been found for all major problem areas. In most cases, these engineering solutions have already been reduced to practical design. Frequently, this was backed up by solid evidence.[99]

For instance, Bell had investigated three solutions to reentry heating. Two ideas were quickly discounted: a hot structure similar to the one developed for the X-15, and a sweat cooling transpiration system that required too much fluid to be practical. The third idea, the same as proposed for BoMi, was an active cooling system that consisted of an aluminum inner wall that carried the structural load, a layer of Dyna-Flex insulation, and a heat-resistant alloy outer skin that could withstand high temperatures but carried no load. Pumps would circulate liquid lithium, liquid sodium, or water between the inner and outer walls and expend vapor overboard. Surprisingly, Bell estimated that the weight of this structure, including the necessary coolant, was comparable to a heat-sink structure constructed of aluminum or titanium alloy. The active cooling system had been tested extensively in the laboratory, and Bell believed it was technically feasible for large-scale applications. During these tests, a liquid-sodium cooling loop for the wing leading edge operated for over 600 hours (total, not continuously) at 1,200 °F, and silicon carbide leading-edge samples had been successfully tested to 4,000 °F. In addition, a 2-foot square sample of the water-wall skin structure had been successfully tested for 90 minutes under full-intensity heating.[100]

Researchers at the Arnold Engineering Development Center had evaluated the proposed nose shape at Mach 16 and at temperatures as high as 10,000 °F. The glider configuration had been tested in a NACA Ames hypersonic wind tunnel up to Mach 10, and various pieces had been tested above Mach 12 using sounding rockets. Panel flutter had been investigated in a NACA Langley wind tunnel at Mach 4 and at temperatures up to 850 °F. Significant test-facility time was being expended on the boost-glide concept.

Bell was still not content with the suborbital approach and believed a spaceworthy BoMi could become the first recoverable piloted satellite. The S-BoMi, as it was called, was expected to achieve an altitude of 480,000 feet (90 miles) at a velocity of 25,640 ft/sec. Bell engineers believed that their knowledge

99. Walter Dornberger, "An Approach to Manned Orbital Flight," presentation to Air Force Scientific Advisory Board, July 29–30, 1957.

100. Ibid.

was sufficient to allow them to develop the suborbital BoMi within 2 years but noted several areas required further study for the orbital S-BoMi. These included the continued development of materials, structural insulation, and cooling techniques, followed by the development of orbital-navigation equipment. Bell also expected that additional aerodynamic problems would result from the interaction of the vehicle skin and the atmosphere in its free molecular state—the composition of the atmosphere at the 50-mile altitude proposed for BoMi was relatively well understood but at 100 miles was essentially unknown. This might have been the most important finding from the various BoMi and RoBo studies; nobody truly understood the environment that the vehicles were expected to operate in nor the technologies needed to survive in it.[101]

To correct this, on November 14, 1956, the Air Force issued SR-131 for the Hypersonic Weapons and Research and Development Supporting Systems (HYWARDS) program, also known as System 610A. The intent of the larger program was to provide research data on aerodynamic, structural, human factors, and component problems associated with high-speed atmospheric flight and reentry in support of BoMi and RoBo. The HYWARDS vehicle was designated System 455L and was primarily intended to serve as a test vehicle for subsystems to be employed in future boost-glide weapons systems.[102]

In support of the Air Force, researchers at NACA Ames and Langley investigated possible configurations for the HYWARDS vehicle. The Langley effort was led by Becker, and the January 17, 1957, report contained the surprising recommendation that the design speed of HYWARDS should be increased to 18,000 ft/sec and at a somewhat lower altitude. Becker pointed out that this was where boost-glide vehicles approached their maximum heating environment, mainly because the rapidly increasing altitudes necessary for speeds above 18,000 ft/sec caused a reduction in heating rates, and the heating rate in space is negligible. An analysis conducted by Langley's Peter F. Korycinski suggested major advantages for a configuration having a flat-bottom surface for the delta wing, with the fuselage located in the relative cool of the shielded area on the leeside (top) of the wing. This flat-bottom design had the smallest critical-heating area for a given wing loading, thus reducing the amount of thermal protection needed. In this respect, the configuration differed considerably from the earlier Bell designs that had used a mid-mounted wing. This was one

---

101. Ibid.

102. System Requirement 131, Hypersonic Weapons Research and Development Supporting Systems, November 14, 1956; Research and Development Project Card, System 455L, December 18, 1956. The "L" suffix was generally applied to reconnaissance vehicles, implying the Air Force may have expected to build some type of operational follow-on vehicle.

of the first clear indications that aerodynamic-design choices could significantly alleviate some of the heating and structural concerns associated with entry. Interestingly, researchers at Ames reached exactly the opposite conclusions.[103]

The Ames HYWARDS configuration was devised by a team that included Allen and Eggers. The configuration presented in a January 1957 report was a flat-top wing body with drooped wingtips that used a combination of surface radiation and an internal-cooling system to control aerodynamic heating. Ames suggested that the "vehicle appears to pose difficult but solvable structural problems." The vehicle had a maximum speed of approximately Mach 10 but a range of only 2,000 miles, compared to 3,300 miles for the Langley design. To produce the highest possible L/D, the Ames design made use of the favorable interference lift that occurred when the pressure field of the underslung fuselage impinged on a high-mounted wing. Unfortunately, the entire fuselage was located in the hottest region of the flow field and the additional weight required to keep the airframe cool outweighed the higher L/D. This would be addressed further in a September 1957 report.[104]

In the September report, Ames discussed two configurations, known as A and B; each could carry a single pilot, his support equipment, and 1,200 pounds of research instrumentation. Both configurations had relatively low wing loadings ($\approx$20 pounds per square foot), used a single XLR99 rocket engine, and required large external boosters to attain the 18,000-ft/sec design velocity. The airframe was similar for both configurations, with a structure that was thermally insulated from the outer skin and actively cooled where necessary.[105]

Research at Ames suggested that the region of highest heating was generally on the high-pressure side of the vehicle. In the A configuration, the fuselage was removed from the high-pressure region by locating it above the wing, which acted as a heat shield, much like with the Langley design. The lower surface of the wing was essentially flat in order to facilitate prediction of pressure distributions and heating rates. The shielding effect of the wing kept the fuselage relatively cool, greatly simplifying the problem of insulating the pilot and instrumentation, and an active cooling system was not required.

Configuration B was essentially the same vehicle proposed in the January 1957 report. This design combined the flow field of a slender expanding fuselage

---

103. Telephone conversations between the author and John V. Becker, various dates in 2001 and 2002.

104. "Preliminary Investigation of a New Research Airplane for Exploring the Problems of Efficient Hypersonic Flight"; "Study of the Feasibility of a Hypersonic Research Airplane," draft NACA Ames Aeronautical Laboratory report (no number) (September 3, 1957).

105. "Study of the Feasibility of a Hypersonic Research Airplane," pp. 6–11.

with the lower surface flow field of the wing to achieve a higher hypersonic L/D at a lower angle of attack, which designers expected would minimize the effects of heating due to generally slower speeds during reentry. Locating the fuselage below the wing also cambered the configuration, making it essentially self-trimming at the maximum L/D, thereby reducing drag due to trimming effects. The drooped, toe-in wingtips were used to achieve static stability and control at hypersonic speeds, supplemented by a retractable ventral stabilizer to improve stability below Mach 6.

The Ames study was notable in that it delved deeply into various thermal protection schemes. Three approaches were investigated in detail: (1) an unprotected hot structure, (2) an internally cooled structure, and (3) an airframe protected from the heat by external insulation. Although the last two approaches were expected to be more complex, the potential weight savings intrigued researchers.

Since Inconel X was being used in the X-15 hot structure, Ames concentrated on the more-advanced alloys that had recently become available. A cobalt-based L-605 alloy demonstrated adequate mechanical properties (it did not become brittle) and oxidation resistance (it did not ablate) up to about 1,800 °F. Some molybdenum alloys retained good strength characteristics even above this temperature but oxidized rapidly without protective coatings, which did not exist at the time. Ceramic materials were not seriously considered for structural elements since they tended to be brittle and had minimal load-carrying abilities, but they were considered essential for small, hot areas such as the wing leading edge.[106]

An investigation of active cooling systems demonstrated the ability to keep temperatures within existing hot-structure technology but left little hope that a workable vehicle could be designed around them. The problem was that coolant is heavy, and even for the small research aircraft being contemplated, over 2,000 pounds of coolant—water or liquid helium—would be needed. The study eventually concludes that "unless high heat-capacity coolants can also be used efficiently for propulsion, the weight of a design in which surfaces are directly cooled appears prohibitive."[107] Since the boost-glide designs were, as the name implies, gliders for most of their flights, even this approach did not hold much promise.

Preliminary investigations into insulating the airframe from heat effects looked promising, but researchers realized that any protective layer must also be capable of withstanding high-speed aerodynamic effects and anticipated

---

106. "Study of the Feasibility of a Hypersonic Research Airplane," pp. 11–17.
107. Ibid., p. 18.

acoustic levels of over 170 decibels. The most promising design for temperatures in excess of 1,800 °F was a metallic superalloy outer skin with closely spaced stand-offs that allowed a thick layer of Thermoflex insulation between the outer skin and the titanium structure underneath. Researchers estimated that this would add about 2 pounds per square foot to the weight of the vehicle. Because of differences in thermal expansion between the insulating and primary structures, expansion joints would be required at appropriate intervals. These, in turn, could lead to surface irregularities that could significantly increase local-heating rates at high speed (due to boundary layer interaction), and they would have to be designed and maintained carefully. The study notes that the "successful development of a smooth and lightweight insulating structure with adequate life expectancy is believed to represent a major development effort." Truer words were never written, but the general approach evolved into the thermal protection system used on the Space Shuttle orbiters.[108]

The September Ames study had added the low-wing A configuration largely in response to criticisms from Langley regarding the initial high-wing configuration proposed in January. Nevertheless, researchers at Ames continued to champion the high-wing design, arguing that, "it is apparent that the flat-top arrangement is the more efficient, claiming tests showed the design offered a 35 percent increase in performance."[109] This design brought to a head the debate within the NACA on the relative merits of flat-top versus flat-bottom designs, a battle that the flat-bottom advocates eventually won. However, while the Air Force and the NACA tried to sort out exactly what HYWARDS should look like, the entire effort was soon overcome by other events.

The Stillborn Round Three X-Plane

In mid-1955, the Soviet Union and the United States had separately announced intentions to launch Earth-orbital satellites as part of the 1957 International Geophysical Year. Nevertheless, when the Soviet Union launched Sputnik on October 4, 1957, the event created a stir among the popular press. The perceived lack of response by President Dwight D. Eisenhower further antagonized the media and, shortly after, the American people, as well. However, it

---

108. "Study of the Feasibility of a Hypersonic Research Airplane," pp. 18–19.

109. Ames Aeronautical Laboratory, "Preliminary Investigation of a New Research Airplane for Exploring the Problems of Efficient Hypersonic Flight," NACA, Washington, DC, January 18, 1957, p. 32.

was the 1,100-pound Sputnik II that ultimately caused the administration to take action since it graphically portrayed the heavy-lift capability of Soviet launch vehicles, and indirectly, of their ICBM program.

By the end of August 1957, the Air Force had already consolidated the BoMi, RoBo, and HYWARDS efforts into a single three-step development program designated System 464L and named Dyna-Soar (for dynamic soaring, as Sänger had referred to skipping reentry). The name was ironic since the skip-gliding concept had been abandoned as unworkable 2 years earlier. The classified title of the new program was "Hypersonic Glide Rocket Weapons System," and by the end of the year, government and industry had spent a little over $8 million ($93 million in 2010 inflation-adjusted dollars) studying the boost-glide concept.[110]

Independently, the NACA Hypersonic Research Steering Committee met on October 15, 1957, to determine the direction of the Round Three research airplane. All agreed that whatever came after X-15 should have a large increment in performance, but no specific requirements were discussed. Perhaps understandably given the launch of Sputnik I, the meeting concentrated more on the future of human space flight, with three different approaches being proposed. A minority led by Faget from NACA Langley argued for a ballistic blunt-body shape—essentially what became the Mercury capsule. Another minority favored the lifting body approach of tailoring a blunt reentry shape to provide a modest L/D that permitted limited maneuvering.[111] The remainder of the committee preferred the flat-bottom hypersonic glider described by Becker. The proposal from Becker appeared to mesh nicely with the new Dyna-Soar effort, so the committee recommended the Round Three requirements be included in the Air Force program.[112]

Step I of the three-step Dyna-Soar program would include a small single-seat hypersonic boost-glide conceptual test vehicle that would provide

---

110. R&D Project Card, System 464L, August 23, 1957, p. 1. Dyna-Soar also stood, apparently, for Dynamic Soarer, as the Air Force called the vehicle itself.

111. Interestingly, later NASA capsule designs had substantial increases in their L/D. In fact, Apollo, with a hypersonic L/D between 0.6 and 0.8, rivaled some of the lifting bodies in cross-range capability, although it was never used operationally.

112. John W. Crowley to Milton Ames, "Meeting of Round III Steering Committee to be held at NACA Headquarters, July 2, 1957" (June 18, 1957); memorandum to the Director from Clotaire Wood, "Presentation to Air Force Headquarters on Round III" (July 11, 1957); and John W. Crowley, NACA Associate Director for Research, to NACA field centers, "Round Three Conclusions" (November 15, 1957), all in possession of authors; "This is Dyna-Soar," Boeing News Release S-6826, 1963.

information on aerodynamic, structural, human factors, and component problems and serve as a test bed for subsystems development. After preliminary air-drop flights, a single-stage launch vehicle would boost the glider to 12,000 ft/sec and 360,000 feet in altitude sometime in 1966. This phase of Dyna-Soar largely satisfied the anticipated Round Three research airplane goals.[113] Step II included a vehicle that could provide high-quality photographic and radar intelligence as well as perform limited bombardment missions. A two-stage booster would propel the vehicle to 22,000 ft/sec at an altitude of 170,000 feet, enabling it to glide 5,750 miles. It was expected that the engines and guidance systems being developed for the ballistic missile programs would be used beginning in 1969.[114] Step III, expected to be operational in 1974, incorporated most of the strategic reconnaissance and bombardment capabilities previously envisioned for RoBo.[115]

By January 25, 1958, the Air Force had screened a list of 111 potential bidders for the conceptual test vehicle, and 10 companies were selected to receive requests for proposals, including Bell, Boeing, Chance Vought, Convair, Douglas, General Electric, Lockheed, Martin, North American, and Western Electric. Three additional contractors—McDonnell, Northrop, and Republic—were subsequently added to the list, although several teaming agreements resulted in only nine actual contenders for the contract.[116] After reviewing the proposals, on June 16, 1958, the Air Force awarded two-phase predevelopment contracts to Boeing-Vought and Martin-Bell to refine their concepts.[117]

X-20 Dyna-Soar

Perhaps the most interesting of all discussions about reentry and recovery from space concerning a winged, reusable vehicle in the 1960s involves the storied career of the X-20 Dyna-Soar. It is one of those flying vehicles about which

---

113. R&D Project Card, System 464L, August 23, 1957, pp. 2–3.

114. Intercontinental missile programs does not necessarily mean ICBMs since Navaho was also considered a prime source.

115. R&D Project Card, System 464L, August 23, 1957, pp. 6–7.

116. Colonel H.M. Harlow to Source Selection Board, "Report of Working Group for System 464L" (January 25, 1958).

117. Major General Ralph P. Swofford, Jr., Acting DCS/Development, Headquarters USAF to Commander ARDC, "Selection of Contractor for WS-464L (Dyna-Soar) Development" (June 25, 1958).

The U.S. Air Force transformed the 1930s spaceplane concepts of Eugen Sänger and others into an experimental vehicle: the Dyna-Soar (Dynamic Soaring). Had it flown, it might have been useful for reconnaissance and long-range bombing from the edge of space. Research on the piloted spaceplane lasted from 1957 to 1963, when the program was canceled before the first vehicle ever flew. The X-20 program was the first serious effort to build and test a winged vehicle that could fly to and from space. USAF.

legend and myth abound.[118] Dyna-Soar was officially designated System 620A on November 9, 1959, and the Boeing-Vought team was declared the winner on December 11, 1959.[119] At the same time, the Glenn L. Martin Company was selected to develop a human-rated version of the Titan launch vehicle.[120] After several reviews and much political infighting, on April 27, 1960, the Air Force awarded a contract to Boeing as the overall integration contractor and ordered 10 Dyna-Soar gliders.[121] The procurement schedule called for two vehicles to be delivered during 1965, four in 1966, and two during 1967. The other two airframes were to be used for static tests beginning in 1965.[122]

The development of Dyna-Soar ranks as perhaps the most convoluted process in modern defense history. Constant fighting between the Air Force and the Department of Defense, particularly Secretary Robert S. McNamara, caused the program to change direction countless times and hindered any real sense of progress. On the other hand, it is a tribute to the technical sides of the Air Force and Boeing that the development effort made real progress. Dyna-Soar was the first serious attempt to build a piloted, reusable, lifting reentry spacecraft, and it began before Project Mercury had taken flight. Most in the industry considered the capsule concept crude, while Dyna-Soar was perceived as elegant since it was maneuverable, reusable, and landed on a runway like an airplane. It was also vastly more complicated and required substantially more research, particularly in the area of thermal protection systems. An entire book would be needed to tell the story, so we will skip most of it since it is beyond

---

118. The most thorough of all studies on this subject is Roy A. Houchin, *U.S. Hypersonic Research and Development: The Rise and Fall of Dyna-Soar, 1944–1963* (New York: Routledge, 2006).

119. This was contract AF33(600)-39831. Vought's involvement with the Dyna-Soar program eventually dwindled to the design and fabrication of the high-temperature nose-cap. This work would later serve Vought (later LTV) well when asked to develop the carbon-carbon nose cap and wing leading edges for the Space Shuttle.

120. This was well before the Titan II was selected for the NASA Gemini program. The letter contract for the Titan II GLV was issued by the Air Force Space Systems Division on January 19, 1962. The Glenn L. Martin Company merged with the American-Marietta Corporation in 1961 to form the Martin Marietta Corporation.

121. The Boeing contract was AF33(657)-7132, and the gliders were assigned serial numbers 61-2374 through 2383.

122. Geiger, "Strangled Infant," pp. 227–226; TWX, RDZSXB-31253-E, Headquarters ARDC to Headquarters Ballistic Missile Division, November 10, 1959; TWX, AFDAT-90938, Headquarters USAF to Commander, ARDC, November 17, 1959.

The launch vehicle for the X-20 evolved considerably over the course of the program, beginning with a Titan II (shown) and ending with a human-rated version of the Titan III. USAF.

the scope of this monograph and concentrate on the aspects of Dyna-Soar related to reentry and recovery from space.

Despite the funding issues, political infighting, and confusion over roles, by the summer of 1961, Boeing had made significant progress on the design of the Model 844-2050 Dyna-Soar I glider. The development effort was directed by Boeing Vice President George M. Snyder in Seattle, WA. The final Dyna-Soar glider was the result of over 13,000 hours of wind tunnel tests that included approximately 1,800 hours of subsonic, 2,700 hours of supersonic, and 8,500 hours of hypersonic time. Several innovative test techniques were used, including a temperature-sensitive paint that was applied to wind tunnel models to determine the heating levels experienced during the test. Dyna-Soar was among the first users of this new paint, which allowed researchers to use relatively inexpensive wood models, instead of metal models, equipped with a myriad of temperature transducers. In addition, a great deal was learned about material science, including new techniques for the manufacture, welding, and extruding of various high-strength superalloys.[123]

The Air Force and NASA reviewed a full-scale mockup on September 11, 1961. The Model 844-2050 was 35.3 feet long, spanned 20.4 feet, and had approximately 345 square feet of wing area. The glider weighed 10,830 pounds empty, and it could carry a maximum payload of 1,000 pounds in a 75-cubic-foot compartment that was pressurized at 0.7 atmosphere with 100-percent nitrogen.[124]

The wing on early Dyna-Soar configurations used a double-wedge upper surface with a flat bottom that provided good hypersonic stability and was easy to manufacture. Unfortunately, this design required the addition of retractable stabilizers for low-speed flight, so the upper wing surface was modified to improve low-speed handling without compromising hypersonic stability. This modification, however, resulted in some transonic instabilities that were eliminated by adding an aft fuselage ramp on the Model 844-2050E, giving the final Dyna-Soar its distinctive appearance.[125]

---

123. "X-20 Dyna-Soar Information Fact Sheet," January 1963, p. 5; Terry Smith, "Dyna-Soar X-20: A Look at the Hardware and Technology," *Quest: The History of Spaceflight Magazine*, Winter 1994, pp. 23–28; X-20 Program Report, Boeing report D2-80852 (March 1963).

124. Dyna-Soar I Structure Description, Boeing report D2-6909-2 (November 1961); "X-20 Dyna-Soar Information Fact Sheet," p. 5; Smith, "Dyna-Soar X-24," pp. 23–28; X-20 Program Report, Boeing report D2-80852 (March 1963).

125. Memorandum, to the Chief of the Research Division, ARDC, "Development of the Dyna-Soar Configuration" (August 23, 1962).

To handle the thermal stresses during entry, the internal structure of the fuselage differed significantly from conventional aircraft of the period. It consisted of a René 41 truss-space frame with fixed and pinned joints in square and triangular elements that looked similar to bridge construction. There were forward and aft beams on each side of the fuselage, with six cross frames connecting each side. In the aft bay, a superstructure of truss members spanned between the main-beam upper chords to react with the skin-support beam loads and provide a load path for torsional flight loads. Wherever possible, the axially loaded truss members were round, swaged-end tubes with sheet-metal tabs or fusion-welded machined fittings. Swaging the tube ends allowed for compact joints at which up to 11 members intersected at a common point. Members subjected to other than pure axial loads (bending or combined bending and axial) were built-up rectangular tube sections fusion-welded together. The space frame was capable of deforming in response to thermal gradients without introducing major stresses or losing structural rigidity. The radiation-cooled structure was designed to survive four reentries.[126]

The thermal protection system for Dyna-Soar was a major challenge, presaging the problems on the Space Shuttle. The structure was designed to achieve an equilibrium-heating condition during entry by radiating heat back to space. Engineers were concerned about the emissivity of the external skins; they were hoping to achieve 0.80, but even a small reduction had a major impact on the expected internal and external temperatures. Unlike ablative or heat-sink concepts, this design placed no time constraints on entry but could only withstand somewhat milder heat pulses. The wide range of L/Ds (and therefore angle of attack) planned for the glider meant that the thermal protection system for each location had to be designed for the worst-case heating environments. The nose cap and wing leading edges received the highest heating at low angle of attack (high L/D), while the bottom of the wing and control surfaces received their highest heating at high angle of attack (low L/D).[127]

Where temperatures were expected to be less than 1,000 °F, the internal truss structure was covered by skin using a series of René 41 panels that were corrugated to add stiffness and allow expansion during entry heating. On the lower surface of the wing, a series of D-36 columbium panels were attached to the underlying René 41 panels using standoff clips with a layer of Dyna-Flex or

126. "X-20 Dyna-Soar Information Fact Sheet," p. 5; Smith, "Dyna-Soar X-20," pp. 23–28; X-20 Program Report, Boeing report D2-80852 (March 1963).

127. *Testing Lifting Bodies at Edwards*. Note that hypersonic gliders utilized the backside of the L/D curve rather than the frontside, where a lower angle of attack would normally yield a lower L/D.

Micro-Quartz™ insulation sandwiched between them.[128] Each external surface panel overlapped the adjoining panel, allowing adjacent panels to expand and move relative to each other without buckling. Special attention was given to heat leakage at the expansion joints, access panels, landing gear doors, and hinge area at the elevon-control surfaces. The wing leading edge, expected to encounter temperatures up to 3,000 °F, was constructed from segmented TZM, an alloy of titanium, molybdenum, and zirconium.[129]

Oxidation of the refractory metals used in the heat shield could lead to structural failure, so a silicide coating was used on the D-36 columbium and TZM panels. A final layer of synar-silicon carbide applied over the silicide coating gave Dyna-Soar its distinctive black color. Although this solved the oxidation issue, these coatings would have had to be refurbished after each flight.[130]

There were two different nose-cap designs, with the early Boeing concept becoming a contingency design after Vought proved to have a better idea. Vought developed a siliconized graphite structure overlaid with zirconia tiles restrained by zirconia pins. In case of cracks in the structure, the tiles and pins were held in place by platinum-rhodium wire. The nose cap was flexibly mounted to a support ring, which in turn was supported at three locations by a linkage arrangement attached to the glider truss structure. The contingency Boeing design used a single-piece zirconia structure reinforced with platinum-rhodium wire. During the molding process, shaped tiles were cast in the outside surface to allow thermal expansion and to prevent possible cracks from spreading. Each design was capable of withstanding 4,300 °F and was later tested during the ASSET series.[131]

The cockpit windows were the largest designed for a piloted spacecraft at the time, and the forward three windows were covered by a D-36 columbium heat shield during the 2,000 °F entry. The cover was jettisoned as the glider slowed to supersonic speeds to provide better visibility during landing. The single window on each side remained uncovered since they were not subjected to high heating loads. NASA test pilot Neil Armstrong used a Douglas F5D

---

128. Dyna-Flex is also known as Cerachrome, while Micro-Quartz goes by the name Q-Fiber Felt. Both are fibrous batt insulating materials that continue to be used in a variety of high-temperature applications.

129. System Development Plan, Dyna-Soar (Step I) Program, 620A, April 1, 1960, pp. III-8 to III-12, III-49 to III-52, III-65 to III-72, III-195 to III-207; System Package Program, System 620A, Dyna-Soar, April 26, 1961, pp. 84–92, 132–139, 182–187.

130. System Package Program, System 620A, Dyna-Soar, April 26, 1961, pp. 84–92, 132–139, 182–187.

131. *Testing Lifting Bodies at Edwards.*

Dyna-Soar was to employ a refractory metal heat shield that engineers expected to use multiple times with only limited refurbishment. This differed significantly from the ablative heat shields used by the capsules and the ceramic heat shield used by the later Space Shuttle. USAF.

Skylancer at Edwards AFB to demonstrate a pilot could land with side vision only in the event that the heat shield did not jettison as planned.[132]

Unlike the capsules, Dyna-Soar was designed to land on a concrete runway 8,000 feet long and 150 feet wide using a three-point landing-skid arrangement based partially on X-15 experience. The Goodyear main landing skids used René 41 bristles wound around a series of longitudinal rods, and the Bendix nose skid (called a dish or a plate) was a single-piece René 41 forging. As it turned out, the landing gear was one of the larger unknowns as the program progressed.[133]

Dyna-Soar had a cross range of 2,000 miles, and the glider was designed to fly at any hypersonic L/D between 0.6 and 1.8, corresponding to angles of attack between 55 and 18 degrees. By combining different bank angles with different L/Ds, the vehicle could fly a wide variety of entry trajectories. Selecting a low L/D with zero bank angle would result in a short, straight-ahead trajectory, while selecting a high L/D with a large bank angle would produce a long, turning entry

---

132. Six pilots had already been selected to fly Dyna-Soar—Captain Albert H. Crews, Major Henry C. Gordon, Captain William J. Pete Knight, Major Russell L. Rogers, Major James Wood (chief pilot), and NASA test pilot Milton O. Thompson. NASA research pilots Neil A. Armstrong and William H. Dana also supported development of the Dyna-Soar.

133. *Testing Lifting Bodies at Edwards*; X-20 Program Report, Boeing report D2-80852 (March 1963).

and high cross range. The footprint for Dyna-Soar at the entry interface (400,000 feet) was approximately 3,000 miles wide and 8,000 miles long, with the size of the footprint getting smaller as the vehicle decelerated. The glider could land anywhere within the footprint. The subsonic L/D was 4.25, and the glider used approach and landing techniques similar to the X-15.[134]

Unfortunately, the end of Dyna-Soar was as convoluted as the beginning. By the summer of 1963, those on the program knew the end was near, but the machinations are well outside the scope of this monograph. When it became obvious that McNamara wanted to cancel Dyna-Soar in favor of a new program suggested by Dr. Harold Brown, Director of Defense for Research and Engineering, NASA made it clear that the Agency did not agree. Dr. Raymond L. Bisplinghoff, The NASA's Associate Administrator for Advanced Research and Technology, pointed out that advanced flight-system studies had repeatedly shown the importance of developing a maneuverable lifting reentry vehicle with a high-temperature, radiation-cooled metal structure. Test facilities were unable to adequately simulate this environment, and, consequently, a research vehicle such as Dyna-Soar was necessary. Bisplinghoff reminded the Department of Defense that NASA had always supported the Dyna-Soar program and would probably need to initiate a substitute research program if Dyna-Soar were cancelled, effectively negating any cost savings to the U.S. Treasury.[135]

On December 4, 1963, Dr. Alexander H. Flax, the Assistant Secretary of the Air Force for Research and Development, wrote to McNamara firmly supporting the continued development of Dyna-Soar and rebutting Brown's argument against the program. Since $410 million had already been expended on Dyna-Soar, Flax questioned the proposal to cancel Dyna-Soar and initiate a new program with similar objectives.[136] The Secretary of the Air Force, Eugene M. Zuckert, and Under Secretary of the Air Force, Dr. Brockway McMillan, agreed with Flax.[137] Zuckert further stated that he did not wish to see the Air

---

134. "X-20 Dyna-Soar Information Fact Sheet," p. 5; Smith, "Dyna-Soar X-20," pp. 23–28; X-20 Program Report, Boeing report D2-80852, March 1963; *General Testing and Ground Support, and Subsystems*, Air Force report ASD-TDR-63-148, March 1963; *Testing Lifting Bodies at Edwards*.

135. R.L. Bisplinghoff, NASA Associate Administrator, to Office of the Secretary of Defense, "X-20 Program," November 22, 1963.

136. Secretary of the Air Force to the Secretary of Defense, "Manned Military Space Program," December 4, 1963.

137. Perhaps significantly, Brockway McMillan was also the Director of the National Reconnaissance Office (NRO), although even the existence of that agency was highly classified at the time. Given that MOL was intended as a reconnaissance platform, it seems odd that the NRO director would oppose the program.

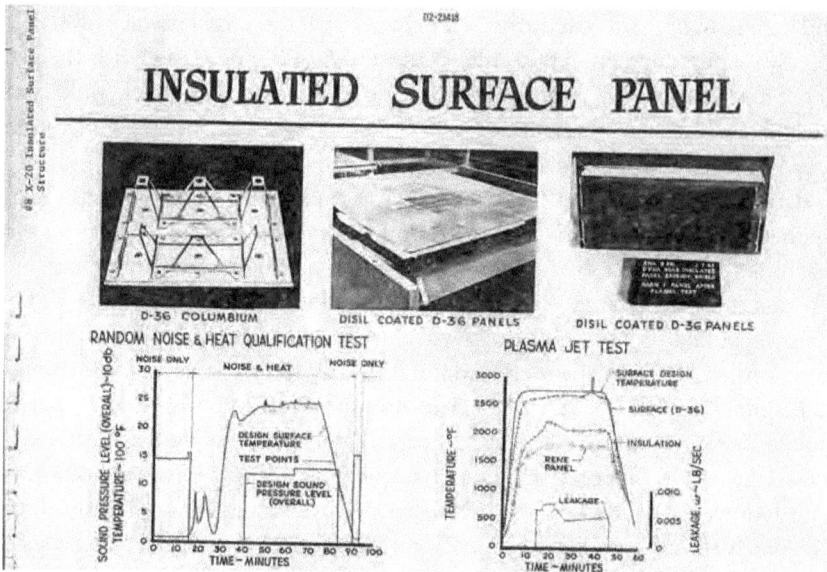

Each of the Dyna-Soar glider's external surface panels overlapped the adjoining panel, allowing adjacent panels to expand and move relative to each other without buckling. USAF.

Force abandon Dyna-Soar and start a new program that had been projected based on undoubtedly optimistic costs and schedules.[138]

It was for naught. On December 10, 1963, McNamara cancelled the Dyna-Soar program. In its place, the Air Force would begin developing the Manned Orbiting Laboratory (MOL) serviced by Gemini-B capsules.[139] The Secretary of Defense explained that Dyna-Soar was not designed to perform space-

---

138. ASAF/R&D to the Secretary of the Air Force, "Manned Military Space Program," December 4, 1963; Secretary of the Air Force to the Secretary of Defense, "Manned Military Space Program," December 4, 1963.

139. The MOL program itself would also be canceled before its first flight. The launch complex (SLC-6) constructed at Vandenberg AFB to support the MOL program was later modified to serve as the west coast Space Shuttle launch site. The STS-33/51-L Challenger accident caused the Air Force to mothball and later cancel plans to use the launch complex. During 1990, it was proposed to rebuild SLC-6 to support Titan IV, a variation of the booster originally intended for Dyna-Soar and MOL. This was also cancelled prior to completion. Over $7 billion was spent on the facility to support various projects, yet the only launches to take place from it were several small Lockheed Launch Vehicle boosters. During 2003–2005, the complex was modified by Boeing to support the Delta IV EELV program. The first successful launch of a large booster was a Delta IV Medium on June 27, 2006.

logistics operations, place substantial payloads into space, or fulfill extended orbital missions. Somehow, McNamara completely ignored what the program had intended to accomplish and criticized it for not being things it had never aspired to. This ended the first serious attempt to build a reusable piloted lifting reentry vehicle.[140]

On December 18, Zuckert approved the continuation of 10 Dyna-Soar activities, including the fabrication of the thermal protection system, the construction of the full pressure suit, the development of the nose cap, and the flight-testing of the coated molybdenum panels and nose caps on ASSET.[141] At the time of cancellation, nearly all of the engineering drawings for the X-20 glider had been released and all of the basic material had been delivered. The construction of the crew compartment was essentially complete and the wing spars, vertical stabilizer spars, and fuselage primary structure were in the final assembly jig. Also at that time, Boeing had 6,475 people involved in the X-20 program, while Minneapolis-Honeywell had 630 and RCA had 565.[142]

Despite this apparent progress, there were several known technical problems looming, although none were considered major hurdles at the time. For instance, Boeing was developing production methods for forming, fastening, drilling, and otherwise working with coated columbium, coated molybdenum, and René 41, and this learning process was causing some delays in the assembly of the glider. Tests had shown that the landing-gear brushes and nose dish were not suited for landings on concrete runways due to excessive wear and

---

140. TWX, AFCVC-1918/63, Headquarters USAF to All Commands, December 10, 1963; News Briefing, Secretary of Defense, "Cancellation of the X-20 Program," December 10, 1963.

141. TWX, ASZR-10-12-1011, Headquarters Aeronautical Systems Division (ASD) to Headquarters Space Systems Division, December 10, 1963; TWX, ASZRK-12-10-249, Headquarters ASD to Air Force Plant Representative (AFPR), Boeing, December 10, 1963; TWX, SAF to Headquarters AFSC, December 11, 1963; Secretary of the Air Force to the Air Force Chief of Staff, "Dyna-Soar Termination," December 12, 1963; X-20 Phase-out Plan, X-20 SPO, December 13, 1963, pp. IV-I to IV-II; TWX, AFRDD-79094, CS/AF to Headquarters AFSC, December 18, 1963.

142. P.J. DiSalvo, Dep. Dir/Procurement, X-20 SPO to AFPR, Boeing Co., "AF33(657)-7132, The Boeing Company Percentage of Completion and SPO Recommendations for Final Settlement," March 12, 1964; DiSalvo to AFPR, Minneapolis-Honeywell Regulator Co., "AF33(657)-7133, Minneapolis-Honeywell Regulator Company Percentage of Completion and SPO Recommendation for Final Settlement," February 27, 1964; DiSalvo to AFPR, RCA, "Contract AF33(657)-7134, Radio Corporation of America, Percentage of Completion and SPO Recommendation for Final Settlement," March 30, 1964; X-20 Detailed Termination Plan, X-20 SPO, January 23, 1964, p. III-2.

inadequate tracking stability. Alternate concepts were under development, but in the meantime, the glider would land on the dry lakebeds at Edwards, just like the X-15.[143]

Several relatively minor structural problems existed at the interface between the glider and the booster, but there was a much more serious problem related to the aerodynamics of the mated vehicle. The winged surfaces of the glider, mounted far forward on the launch vehicle, created instabilities during the early boost phase.[144] The addition of fins at the bottom of the Titan III booster (as proposed on an earlier Titan II version) regained the static stability, but wind tunnel tests uncovered some unexpected aerodynamic interference between the forward wing and the aft fins, largely negating their effectiveness. The added weight of the fins was also undesirable. Increasing the control authority of the rocket nozzles allowed the instability to be properly controlled without the fins, but this introduced excessive structural loads in the booster and at the glider/booster attachment point while transiting wind shears early in the ascent. At the program's termination, a solution was still being sought.[145]

When Dyna-Soar was canceled, the first X-20 air launch from the B-52C was about 18 months out, and the first flight on a Titan IIIC was 30 months in the future.[146] Approximately $410 million ($3.7 billion in 2009 inflation-adjusted dollars) had already been spent, and another $373 million was needed prior to the first orbital flight. Even if Dyna-Soar had never flown an operational mission, it would have provided valuable information on entry flight control and heating, something that was seriously lacking during the development of the Space Shuttle 10 years later.

The 1970 Hypersonic Vehicles Report: Setting the Stage for the Future

Test models of many of the concepts were flown in wind tunnels, and a few in free flights. For instance, in 1969 North American built a 450-pound model of an elongated, conical piloted reentry vehicle that had a set of rotors stowed in the upper fuselage. As the vehicle slowed to subsonic speeds, the rotors deployed and the blades began to unfold. As speed reduced further, the rotors

---

143. *Testing Lifting Bodies at Edwards.*

144. The lifting body concepts eliminated, or at least minimized, this destabilizing effect since the overall vehicle width would have been about the same as the diameter of the booster.

145. *Testing Lifting Bodies at Edwards.*

146. Ibid.

opened fully and the vehicle autorotated until it reached the landing area. The tests demonstrated that the concept worked, and the weight penalty was not as extreme as first expected. However, design studies indicated that available technology would limit the application of the technique to vehicles weighing less than 50,000 pounds. This was unacceptable for projected programs such as the Space Shuttle.[147]

147. E.P. Smith, "Space Shuttle in Perspective: History in the Making," AIAA paper 75-336 presented at the AIAA 11th Annual Meeting and Technical Display, Washington, DC, February 24–26, 1975, p. 6.

When the concept of a space shuttle emerged in the latter 1960s, NASA envisioned it as a two-stage, fully reusable vehicle with both parts landing on a runway. This wind tunnel model represented the "DC-3" concept developed by Max Faget at the Manned Spacecraft Center (renamed the Johnson Space Center in 1973). NASA.

CHAPTER 5

Spaceplane Reality

Although we tend to think of the Space Shuttle as more modern than the capsules that preceded it, such was not truly the case. Dyna-Soar, Mercury, and Apollo were all first-generation spacecraft, and their development took place mostly concurrently, with little chance for one program to benefit greatly from the others. Gemini came later and benefited from Mercury to become the first second-generation spacecraft (although one could argue that the Apollo Block II Command Module also qualifies). Since the development of Dyna-Soar was cancelled before its first flight, the Space Shuttle represents the first-generation lifting reentry spaceplane. When the initial studies that led to the Space Shuttle began, the United States had completed fewer than two-dozen piloted space flights using the three capsule types; indeed, when Columbia made her maiden voyage, the United States had only made 30 orbital flights. It was an extremely small base of experience upon which to begin such an ambitious program. The audacity of the space community in successfully bringing this revolution in space flight to fruition should not be minimized in assessing the history of the Space Shuttle.

The engineers who created the Space Shuttle did so with grand expectations. However, the exercise must be looked at in the context of its time. The United States was riding high on the successes of the early human space flight programs, particularly the Apollo Moon landing program, and visions of space stations and the space shuttles that serviced them were firmly implanted in the minds of the engineers and the public. Arthur C. Clarke and Stanley Kubrick furthered these ideas to the music of "The Blue Danube" in the movie *2001: A Space Odyssey.*

Since the beginning, most visions of human space flight revolved around the concept of a spaceplane, a logical extension of the airplane that had become commonplace during the 20th century. It came as a disappointment to many engineers and researchers when the first humans ventured into space aboard capsules. Test pilots were known to remark that astronauts were "Spam in a can"—many engineers were not much kinder in their critiques of the capsules used by Mercury, Gemini, and Apollo. However, the simple fact was that there was no other way to get humans into space at the time. Redstone and Atlas simply were not powerful enough to launch a spaceplane, and the political realities dictated that the United States could not wait.

Titan probably was powerful enough to launch a small winged vehicle, perhaps like the one proposed by John Becker at the NACA Conference on High-Speed Aerodynamics in March 1958. However, by the time Titan flew, President John F. Kennedy had committed to a Moon landing before the end of the decade, so the momentum for capsules continued. Nevertheless, visions of spaceplanes were abundant, and the Air Force committed to building the first reusable, lifting reentry spacecraft when it began the development of Dyna-Soar. The X-20 effort took little from the NASA capsules since their development was concurrent, and the entire concept was much more advanced. Instead of throwing away the vehicle after every flight, as with the capsules, Dyna-Soar was designed to fly four times with only minor refurbishment of its superalloy metallic heat shield. It would be maneuverable during entry and able to land on a conventional runway. It was also incredibly controversial within the Department of Defense and ultimately was cancelled by Secretary of Defense Robert S. McNamara before the first vehicle was completed. Regardless of whether there was a valid military mission for the spacecraft, its development and flight-test would have provided a much-needed body of knowledge when it came time to build the next winged space vehicle—the Space Shuttle.

NASA and the Air Force began studying vehicles that resembled space shuttles in the late 1950s and continued throughout the 1960s. The development of the actual vehicle called the Space Shuttle began when the NASA Manned Spacecraft Center (MSC) and the Marshall Space Flight Center (MSFC) released a joint Request for Proposals (RFP) for an 8-month study of an Integral Launch and Reentry Vehicle (ILRV). Five companies—the Convair Division of General Dynamics, Lockheed, McDonnell Douglas, Martin Marietta, and North American Rockwell—expressed an interest in participating in the study, which was Phase A of the Space Shuttle design and development cycle.[1] In NASA's four-phase process, Phase A was labeled advanced studies, Phase B was project definition, Phase C was vehicle design, and Phase D was production and operations. The 21st century Constellation program is following a similar process.

Throughout the Phase A studies, there was considerable technical controversy within NASA over the size and configuration of the space shuttle. These studies—supported by more than 200 man-years of engineering effort and backed up by countless hours of wind tunnel testing, material evaluation, and structural design—resulted in four basic vehicle types. These included straight-wing, delta-wing, and stowed-wing vehicles as well as lifting bodies. Combined with independent Air Force ILRV studies, Phase A confirmed that cross range was the major controversy when trying to consolidate Air Force requirements with the

---

1. Request for Proposals (RFP), MSC-BG721-28-9-96C and MSFC-1-7-21-00020, October 30, 1968.

needs of NASA. The major issue was the Air Force wanted the space shuttle to be able to land at its launch site after only a single polar orbit. During this time, Earth had rotated approximately 1,265 miles, meaning the returning spacecraft had to be able to fly at least that distance during entry. On the other hand, NASA simply wanted an opportunity to land back at the launch site from low-inclination orbits once every 24 hours, requiring a relatively modest 265 miles in cross-range capability (although various abort scenarios raised this to almost 500 miles in most studies). Still later, abort considerations would raise the NASA requirement to over 1,000 miles, almost coinciding with the Air Force's requirement. In addition, national-security considerations also drove the decision on the size of the payload bay. The reconnaissance program had a particular satellite being planned for deployment from the Shuttle, and this drove the length of the payload bay (the diameter was driven by the size of the expected modules for a space station).

During Phase A, engineers found the lifting body concept to be the most poorly suited to space shuttle applications, primarily because the shape did not lend itself to efficient packaging and installation of a large payload bay, propellant tanks, and major subsystems. The complex double curvature of the body resulted in a vehicle that would be difficult to fabricate, and further, the body could not easily be divided into subassemblies to simplify manufacturing. In addition, the lifting body's large base area yielded a relatively poor subsonic lift-to-drag ratio (L/D), resulting in a less attractive cruise capability. Although lifting bodies continued to be studied throughout Phase B, the concept was a dark horse, at best.

At the same time, stowed-wing designs were found to have many attractive features, including low burnout weight and the high hypersonic L/D needed to meet the maximum cross-range requirements. In addition, the stowed approach permitted the wing to be optimized for the low-speed flight regime, simplifying landing. Its drawbacks included a high vehicle-weight-to-projected-planform-area ratio, which would result in a higher average base temperature relative to either straight- or delta-wing concepts. In addition, the mechanisms needed to operate the wing and transmit the flight loads to the primary structure resulted in significantly increased design and manufacturing complexity. The maintenance required between flights was expected to be high, and insufficient data existed to reliably determine potential failure modes, which were thought to be numerous.

Maxime A. Faget, having moved from Langley to MSC in Houston, TX, was not a proponent of lifting reentry. Instead, Faget held to the idea of a high-drag blunt body. He designed a straight-wing space shuttle design popularly called the DC-3 (more officially, the MSC-001), which operated as a blunt body. Faget proposed to enter the atmosphere at an extremely high angle of attack with the broad lower surface facing the direction of flight, creating a large shock wave that would carry most of the heat around the vehicle instead of into it. The vehicle would maintain this attitude until it got below 40,000 feet and about 200 mph, when the nose would come down, and it would begin diving to pick up sufficient speed to fly toward the landing site, touching down at about 140 mph. Since the only flying the vehicle would perform was at very low speeds during the landing phase, the wing design could be selected solely on the basis of optimizing it for subsonic cruise and landing; hence the simple straight wing proposed by Faget. The design did have one major failing (at least in the eyes of the Air Force): Since it did not truly fly during reentry, it had almost no cross range.[2]

Faget had convinced many in NASA that his simple straight-wing concept would be more than adequate. Nevertheless, not everybody agreed. In particular, Charles Cosenza and Alfred C. Draper, Jr., at the Air Force Flight Dynamics Laboratory did not accept the idea of building a space shuttle that would come in nose high, then dive to pick up flying speed. For most of the entry, the vehicle would be in a classic stall, and the Air Force disliked both dives and stalls, regarding them as preludes to out-of-control crashes. Draper preferred to have the vehicle enter a glide while still at hypersonic speeds, thus maintaining much better control while still avoiding much of the severe aerodynamic heating.[3] It should be noted that even Draper agreed with the falling concept for the first part of entry to take advantage of the blunt-body theory; the difference lay with exactly when to begin flying.

However, if Draper's vehicle was going to glide across a broad Mach range, from hypersonic to subsonic, it would encounter another aerodynamic problem: a shift in the wing's center of lift. At supersonic speeds, the center of lift is located about midway along the wing's chord (the distance from the leading to the trailing edge); at subsonic speeds, it moves much closer to the leading edge. Keeping an aircraft in balance requires the application of aerodynamic forces that can compensate for this shift. Another MSC design, the Blue Goose, accomplished this by translating the entire wing fore to aft as the center of lift changed—a heavy and complex solution.

Air Force engineers had found that a delta planform readily mitigated most of the problem, so Draper proposed that any space shuttle should use delta wings instead of straight ones. Faget disagreed, pointing out that his design did not fly at any speed other than low subsonic; at other speeds, it fell and was not subject to center-of-lift changes since it was not using lift at all. Faget argued

---

2. Maxime Faget and Milton Silveira, "Fundamental Design Considerations for an Earth-surface-to-Orbit Shuttle," NASA study 70A-44618 (October 1970).

3. T.A. Heppenheimer, *The Space Shuttle Decision: NASA's Search for a Reusable Space Vehicle* (Washington, DC: NASA SP-422, 1999), p. 210.

that a straight wing with a narrow chord would be light and have relatively little area that needed a thermal protection system. To achieve the same landing speed, a delta wing would be large, add considerable weight, and greatly increase the area that required thermal protection.[4]

Draper argued that delta wings had other advantages. Since it was relatively thick where it joins the fuselage, a delta wing offered more room for landing gear and other systems that could be moved out of the fuselage, freeing space for a larger payload bay. Its sharply swept leading edge produced less drag at supersonic speeds, and its center of lift changed slowly compared to a straight wing. In addition, the delta wing offered one other advantage—one that became increasingly important as the military became more interested in using the Space Shuttle. Compared to a straight wing, a delta wing produces considerably more lift at hypersonic speeds. This allowed a returning vehicle to achieve a substantial cross range during entry, a capability highly prized by the Air Force. It was a complicated problem, and one that would take at least another year to solve.

Phase B Studies

To help resolve the controversy, two concepts were baselined for the Phase B study, including a Faget-style straight-wing, low cross-range orbiter, and a Draper-supported delta-wing, high cross-range orbiter. The straight-wing orbiter would be configured to provide design simplicity, low weight, decent handling, and good landing characteristics. The vehicle would be configured to enter at a high angle of attack to minimize heating and facilitate the use of a heat shield constructed from materials available in the early 1970s. The delta-wing orbiter would be designed to provide the capability to trim over a wide angle-of-attack range, allowing initial entry at a high alpha to minimize the severity of the heating environment, then transitioning to a lower angle of attack to achieve a high cross range. Of course, the entry concept for the Space Shuttle was significantly different from that intended for Dyna-Soar. Dyna-Soar was designed to complete an entire entry at any angle of attack from about 18 degrees to 55 degrees. This resulted in over-design of the thermal protection system for those regions that were only critical during the extreme low or high angles of attack. The Space Shuttle was designed for a predefined angle-of-attack schedule with a Mach number that allowed optimization of the thermal protection system but less flexibility in achieving the desired cross range or downrange.

---

4. Ibid., pp. 211–212.

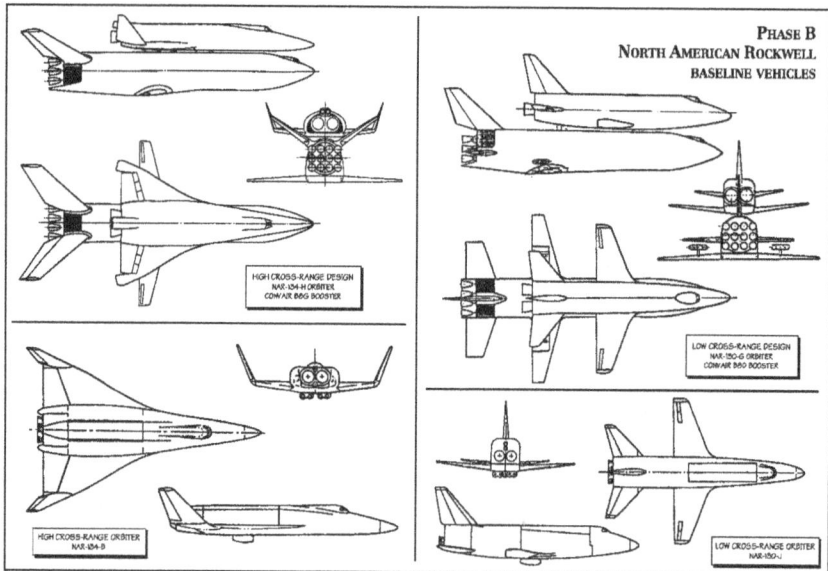

Four design concepts for the Space Shuttle. NASA/Dennis R. Jenkins.

NASA issued two Phase B contracts on July 6, 1970: one to a team consisting of McDonnell Douglas and Martin Marietta, and the other to North American Rockwell, which was joined by Convair as a risk-sharing subcontractor. This phase involved a detailed analysis of mission profiles and a preliminary look at the R&D needed to complete them. The effort would also include a great deal of study on thermal protection systems, a critical component in any spaceplane concept.

The Phase B studies were the first comprehensive review of the thermal protection system for a large lifting reentry vehicle. These studies included extensive investigations into the expected heating environments, and the various materials that might be used on a space shuttle. The earlier Dyna-Soar work was largely overcome by the expected, and unrealistic, high flight rate and the size and weight of the new vehicle.

Any thermal protection system is defined by the natural and induced environments to which it is exposed during various phases of the mission. Natural environments include rain, dust, wind, and micrometeoroids, which may require, in certain instances, that additional protective devices (e.g., coatings) be applied to the basic heat shield. Induced environments include heating, pressures, and acoustics that result from vehicle aerodynamic design and the selected flight profile.[5]

---

5. D.W. Hass, "Final Report: Refurbishment Cost Study of the Thermal Protection System of a Space Shuttle Vehicle," McDonnell Douglas Astronautics Co., Contract NAS1-10093, NASA CR-111832 (March 1, 1971).

One of the most promising ideas to come out of the Phase A studies was to mount the space shuttle thermal protection system so that it could be easily removed from the vehicle. This allowed the systems and structure under the thermal protection system to be serviced and the thermal protection system itself to be refurbished without taking the entire vehicle out of service for a prolonged period. In this concept, whatever material the thermal protection system was made out of would be mounted on panels, and these panels would be installed over the basic vehicle structure with an appropriate insulation between. When the vehicle returned from orbit, the panels would be replaced; the old panels would be sent to a shop for refurbishment, and then used on a future mission. This made sense at the time since it appeared that the most likely materials were ablators and refractory metals, both of which would need significant maintenance between flights.

Ablators were the departure point for most of the thermal protection system studies because they were well understood and proven. Researchers studied various materials, including silica/silica-fiber composites; mixed inorganic or organic composites with silica, nylon, or carbon fiber-reinforced resins (phenolics, epoxies, and silicones); and carbon- or graphite-based materials. These ablators had densities ranging from 10 pounds per cubic foot (using microballoons to keep the material lightweight) to 150 pounds per cubic foot (for high-heat protection). Researchers evaluated both charring and noncharring materials, but the need to maintain a precise outer mold line for aerodynamic efficiency usually mandated charring ablators.

The primary application technique for elastomeric materials was spraying, although the experience of applying MA-25S ablator on the X-15A-2 had been less than ideal. The ablator could be sprayed directly on primary structures and cured (as with the X-15A-2) or sprayed on panels and attached by fasteners. A major problem with spraying an ablator directly onto the vehicle was that most ablators required curing at elevated temperatures, meaning a very large oven would need to be constructed.[6]

One of the bigger unknowns with ablator heat shields was how to refurbish them. Researchers had intended to investigate this aspect of ablators using the X-15A-2 at Edwards AFB, but the first high-speed flight with the ablative coating resulted in substantial damage to the airplane, and further flight tests were cancelled. In truth, the damage was not a failure of the ablator but an engineering flaw that failed to recognize the potential for shock impingement on an experimental ramjet that was hung below the ventral stabilizer. In addition, the X-15 used a hot structure to tolerate the heat on its flights, and the

---

6. Ibid.

ablator interfered with the normal radiance and absorption of heat that the structure depended on. It was not a happy experience.[7]

Refurbishing ablators requires the removal of the used ablator and replacing it with new material. For a panel design, removing the used panels and substituting new ones minimized vehicle turnaround time. The used panels could then be refurbished or discarded. Direct application to the vehicle skin required that the used ablator be removed from the primary structure and new ablator applied, likely increasing vehicle turnaround time. In either case, elastomeric ablator facilitated refurbishment, as it was soft and easily cut with a knife or other sharp edge, layer by layer, down to the substrate.[8]

However, elastomeric ablator refurbishment did not necessarily require the complete removal of the used ablator. It would be possible to apply a thick enough coat of ablator to last for several flights. After the first flight, only the charred or spent ablator would be removed—a relatively simple task. However, this imposed an initial weight penalty. Another approach was to refurbish the ablator by removing only the charred or spent ablator down to virgin material. New ablator could then be applied over the virgin material. This was how the X-15 program intended to refurbish the X-15A-2. The thickness was measured by pins embedded in the ablator at specified intervals, followed by manual sanding until the desired height was attained—a tedious, manual process. Obviously, a better technique would need to be developed.

Dyna-Soar had eschewed ablators in favor of a metallic heat shield, and this seemed an elegant solution to most researchers. The use of radiative metallic panels supported by a cool structural shell with a layer of insulation between would work for all but the hottest regions of the vehicle. Metals are intrinsically durable materials and therefore capable of extensive reuse, but they have oxidation or strength limitations that needed resolution prior to using them in an operational thermal protection system. However, temperature limitations are largely determined by the material's ability to resist oxidation. Unfortunately, most of the exotic and superalloys tested for these programs tended to oxidize quickly when exposed to high temperatures. To minimize this oxidation, researchers developed coatings that could be applied to the metal to shield it from the effects of the atmosphere. At the time, coatings had been developed that appeared to permit an upper-limit temperature of about 2,500 °F for 100 cycles, but actual real-world data was missing.[9]

---

7. Dennis R. Jenkins, *X-15: Extending the Frontiers of Flight* (Washington, DC: NASA SP-2007-562, 2008), pp. 442–466.

8. Hass, "Final Report."

9. Ibid.

A fully reusable Space Shuttle design concept from 1970. NASA.

The most likely candidates for metallic radiative heat shields included titanium, nickel- and cobalt-based superalloys, thoria dispersed nickel-chromium (TD NiCr), and coated columbium alloys. Typically, nickel alloys and cobalt alloys were used in environments in which temperatures exceeded 1,000 °F. Due to their chemical compositions, these materials tended to be quite expensive, with long lead times from the mills. Some of the more well-known trade-marked alloys included Inconel®, Hastelloy®, Monel®, Nitronic®, CBS600, Haynes 188®, L605, and René 41®.

Selecting the upper-use temperature limit for a metallic material involved the evaluation of creep strength, metallurgical stability, rate of oxidation, life of the oxidation-protective coating, and effect of oxidation-protective coatings on the mechanical properties of the base materials. In general, titanium alloys were shown to be effective below 900 °F, and if the panels were exposed to only minor pressure loads, short excursions to environments up to 1,200 °F could be tolerated. For the large surface areas subjected to temperatures between 1,200 °F and 1,800 °F, conventional nickel- and cobalt-based superalloys proved efficient and economical.[10]

Researchers had more difficultly finding suitable material for temperature ranges between 1,800 to 2,200 °F. Conventional superalloys proved structurally inefficient above 1,800 °F, while coated columbium was not weight efficient below 2,200 °F. Researchers believed that TD NiCr alloy was the most promising candidate to fill this temperature range. In addition to being lighter, a major advantage of TD NiCr over columbium alloys was that it did not require protective oxidation coatings. However, at the time, TD NiCr was a relatively new material and very little information was available regarding its use in a thin-gauge reusable heat shield application.[11]

New materials, including hardened compacted fibers (HCF) and oxidation-inhibited carbon-carbon, which were neither ablative nor metallic, presented engineers with an entirely different set of options. The most common HCF investigated during the early studies was called mullite, though it was not truly mullite (a rare chemical that is found only on the Isle of Mull off the west coast of Scotland). Mullite had long been used as a refractory material in the glass and steel industries since it exhibited high temperature strength, good thermal-shock resistance, and excellent thermal stability. Since mullite was rare, industry researched and discovered methods of using various materials to fabricate synthetic mullite ceramics.[12]

HCF was characterized by a layer of rigidized inorganic fibers that combined functions of a high-temperature radiative surface and an efficient insulation. Researchers believed that HCF was a promising candidate for space shuttle applications because of its potential weight savings (it weighs only 12–15 pounds per cubic foot), availability, and temperature-overshoot capability. These HCF materials were relatively soft, extremely porous, and had inherently low emittance values. Adhesive bonding had to be used since HCF material would not support mechanical fasteners due to structural weakness. The materials that were candidates to produce HCF included aluminosilicate, silica, aluminosilicate-chromia, artificial mullite, and zirconia. Researchers believed these materials

---

10. Ibid.

11. Ibid.

12. Mullite fact sheet, located at *http://www.azom.com/details.asp?ArticleID=925*, accessed October 1, 2009.

were potentially useful for 100 flights in the range of 2,000 to 3,000 °F. They also believed that the development of suitable reusable coatings was the most critical problem that had to be solved before the material could be successfully used on any space shuttle. In many ways, the mullite shingles were an indication of things to come as the ceramic Lockheed insulation was perfected.[13]

A relatively new class of materials, all of which benefited significantly from the investment in Dyna-Soar, was also extensively investigated. Researchers believed that oxidation-inhibited, fiber-reinforced, carbonaceous materials offered the potential for a reusable, reasonably cost efficient, high-temperature-resistant structural material. The strength-to-weight ratio of these materials at high temperatures was unsurpassed, leading to its use on the wing leading edge and nose cap, where aerodynamic forces were highest. The carbonaceous material was reinforced by carbon or graphite fibers, usually in the form of cloth. It should be noted that these materials, like the reinforced carbon-carbon actually used on the Space Shuttle, were not insulators, and the backside of the material was essentially as hot as the frontside. What the materials provided was a means of creating the appropriate outer mold line on the hottest parts of the vehicle (the nose and wing leading edge), but some sort of insulating material was still needed to protect the primary structure. The insulating material was usually a fibrous material characterized by low strength, low density, and low thermal conductivity. Insulators typical of this class include Q-Felt®, Micro-Fiber®, and Dynaflex® batt insulation.

Reusable Surface Insulation

However, a new material was just being developed. Although it was similar in many respects to the artificial-mullite shingle investigated during Phase B, this new material represented a quantum leap in technology. During the original studies of lifting reentry vehicles in the late 1950s, there had been a great debate over the relative merits of active cooling systems versus passive systems for thermal protection. The active systems were attractive on paper but nobody could quite figure out how to make them work at a reasonable weight. Therefore, the choices were largely narrowed to either a hot structure, like the X-15, or a more conventional structure protected by some sort of insulation. The hot-structure approach required the use of rare and expensive superalloys, and there was always a great deal of doubt if it would have worked on a vehicle as large as a space shuttle. Generally, most preferred a fairly conventional structure made

---

13. Ibid.

of titanium or aluminum protected by a series of metallic shingles with a thick layer of insulation in between the two. There was some investigation of ablative coatings, but the unhappy X-15A-2 experience made just about everybody shy away from this technology except as a last resort.

Things began to change as the Lockheed Missiles and Space Company made quick progress in the development of a ceramic reusable surface insulation (RSI) concept, and by December 1960 Lockheed had applied for a patent for a reusable insulation material made of ceramic fibers. The first use for the material came in 1962 when Lockheed developed a 32-inch-diameter radome for the Apollo spacecraft made from a filament-wound shell and a lightweight layer of internal insulation cast from short silica fibers. However, the Apollo design changed, and the radome never flew.

Nevertheless, this experience led to the development of a fibrous mat, called Lockheat, that had a controlled porosity and microstructure. The mat was impregnated with organic fillers such as methylmethacylate (plexiglass) to achieve a structural quality. These composites were not ablative, since they did not char to provide protection. Instead, Lockheat evaporated, producing an outward flow of cool gas. By 1965, this had led to the development of LI-1500, the first of what became the Space Shuttle tiles. This material was 89-percent porous, had a density of 15 pounds per cubic foot (hence the 15 in the designation), was capable of surviving repeated cycles to 2,500 °F, and appeared to be truly reusable. A test sample was flown on an Air Force reentry test vehicle during 1968, reaching 2,300 °F with no apparent problems.

Lockheed engineers continued developing the silica RSI, and Lockheed decided to produce the material in two densities to protect different heating regimes: 9 pounds per cubic foot (designated LI-900) and 22 pounds per cubic foot (LI-2200). The ceramic consisted of silica fibers bound together and sintered with other silica fibers, then glaze-coated by a reaction-cured glass consisting of silica, boron oxide, and silicon tetraboride. Since this mixture was not waterproof, a silicon polymer was coated over the undersurface (i.e., unglazed) side. The material was very brittle, with a low coefficient of linear thermal expansion, and therefore Lockheed could not cover an entire vehicle with a single piece of it. Rather, the material needed to be installed in the form of small tiles, generally 6-by-6-inch squares. The tiles would have small gaps between them to permit relative motion and to allow for the deformation of the metal structure under them due to thermal effects. Since a tile would crack as the metal skin under it deformed, engineers decided to bond the tile to a felt pad and then bond the felt pad to the skin. Both of these bonds were accomplished with a room temperature vulcanizing (RTV) adhesive. It was a breakthrough, although everybody involved approached the concept with caution.

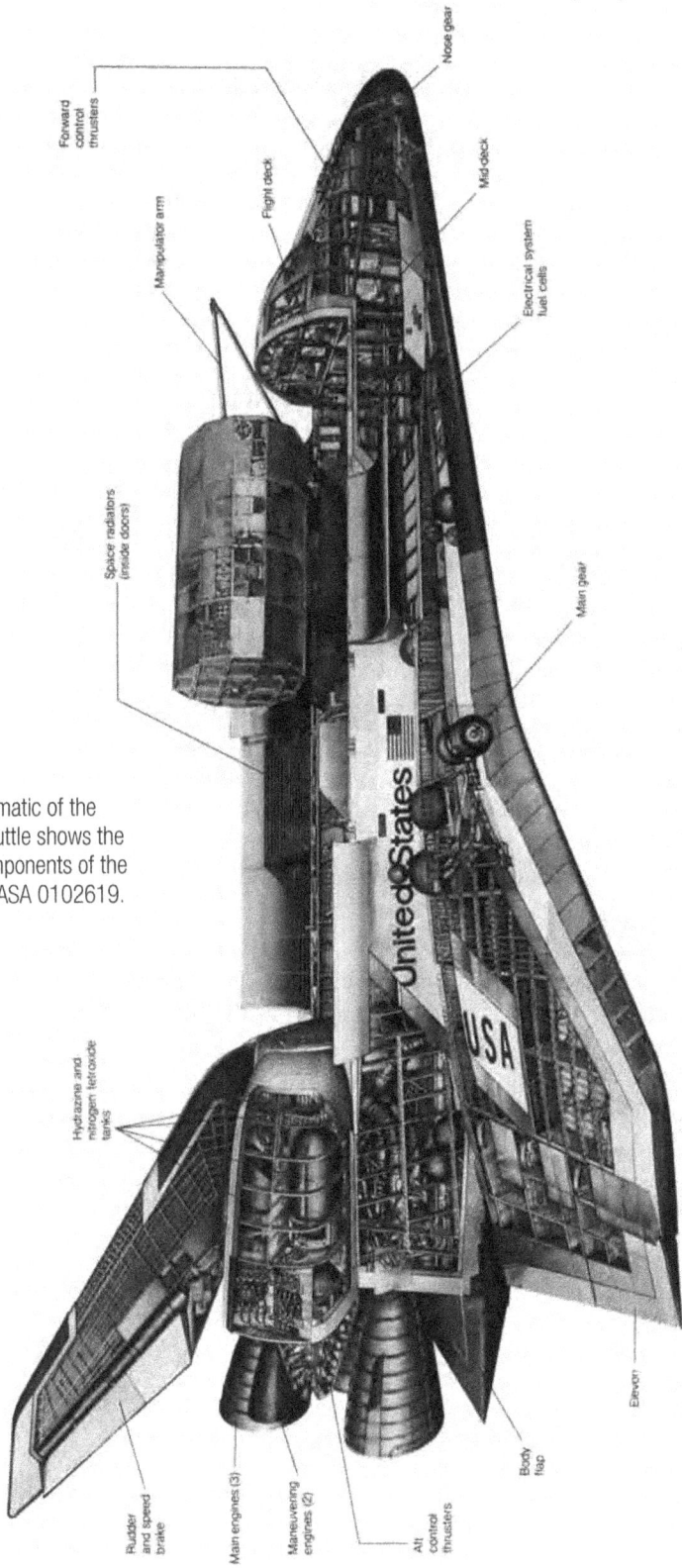

Nose gear

Forward control thrusters

Flight deck

Mid-deck

Manipulator arm

Electrical system fuel cells

Space radiators (inside doors)

Main gear

United States

USA

This schematic of the Space Shuttle shows the major components of the vehicle. NASA 0102619.

Hydrazine and nitrogen tetroxide tanks

Rudder and speed brake

Main engines (3)

Maneuvering engines (2)

Aft control thrusters

Body flap

Elevon

Phase C/D Developments

It appears to be the fate of winged spacecraft that their development is shrouded in political maneuvering and controversy. So it had been with Dyna-Soar, and so it was with the Space Shuttle. After long and heated battles within NASA, the Department of Defense, the White House, the Office of Management and Budget, and Congress, President Richard M. Nixon finally approved the development of the Space Shuttle on January 5, 1972.[14]

The RFP was issued on March 17, 1972, to Grumman/Boeing, Lockheed, McDonnell Douglas/Martin Marietta, and North American Rockwell for the production and initial operations of the Space Shuttle system. The technical proposals were due on May 12, 1972, with the associated cost and contractual data due a week later. By this time, the Faget low cross-range concept had been abandoned, and the Government baseline orbiter, called MSC-040C, had a large delta wing that met all of the Air Force's requirements.[15]

The RFP stated that each orbiter should have a useful life of 10 years and be capable of up to 500 missions. However, the RFP asked each contractor to provide information on the impact of lowering this to only 100 missions, a figure that was subsequently adopted. Three reference missions were specified: (1) lifting 65,000 pounds of payload into a 310-mile, due-east orbit from the Kennedy Space Center (KSC); (2) 25,000 pounds into a 310-mile, 55-degree orbit from KSC while carrying air-breathing landing engines; and (3) 40,000 pounds into a 115-mile, circular polar orbit from Vandenberg AFB, CA. The orbiter was to have the capability to return to its launch site after a single orbit, although an actual cross range was not specified. An unspecified RSI thermal protection system was to be used, although ablative material or other special forms of thermal protection systems were allowed where beneficial to the program. The orbiter thermal protection system was to be capable of surviving an abort scenario from a 500-mile circular orbit.[16]

Unsurprisingly, the four contractor-proposed vehicles looked remarkably similar, since all were based on the MSC-040C concept. Grumman and North American were the only competitors that designed the orbiter for the baseline 500-mission service life; the others opted for the 100-mission alternate.[17]

---

14. Richard M. Nixon, Statement by the President, January 5, 1972, and NASA Press Release 72-4, January 6, 1972, both in NASA Historical Reference Collection.

15. John M. Logsdon, "The Space Shuttle Program: A Policy Failure?" *Science* 232 (May 30, 1986): 1099–1105, especially p. 1102.

16. RFP Space Shuttle Program, pp. III-1 to III-2, IV-10, IV-12.

17. Ibid., p. IV-24.

Wing Structure

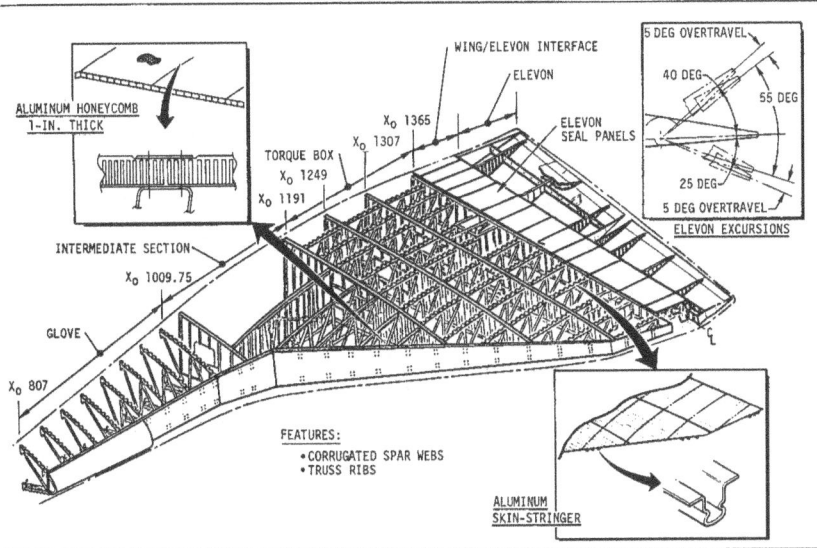

The thermal protection system of the Space Shuttle relied on ceramic tiles that were especially heat resistance but were fragile and required considerable care to ensure their successful use. It also used reinforced carbon-carbon on the wing leading edge and nose cap. This drawing of the orbiter wing shows the structure of this critical component of the vehicle. NASA.

Grumman did not completely trust the Lockheed-developed RSI and proposed adding a layer of ablative material around the crew compartment and the orbital maneuvering system (OMS)/reaction control system (RCS) modules during the early development flights to guard against a possible RSI burn-through. After landing at KSC, the Grumman orbiter would be purged and secured, towed to a new refurbishment facility where the thermal protection system would be repaired as necessary, then sprayed to maintain the thermal protection properties and prevent water absorption.[18]

The Lockheed Phase C/D proposal took some liberties with the basic MSC-040C design. Specifically, the wing planform showed a decided bias toward a double-delta design long before the program officially sanctioned this change. Surprisingly, the proposal did not specifically address why Lockheed had chosen this design, other than it slightly simplified the thermal protection

---

18. Proposal for Space Shuttle Program (multiple volumes), Grumman report 72-74-NAS (Bethpage, New York: Grumman Aerospace Corporation, May 12, 1972).

system. For instance, Lockheed proposed LI-1500 tiles for the inboard part of the leading edge, using the more expensive carbon-carbon system only on the outboard sections. The overall orbiter thermal protection system was designed to provide 100 normal operational entries at 2,500 °F, or a single-contingency reentry at 3,000 °F.[19]

McDonnell Douglas remained fairly true to the original MSC-040C design, but its thermal protection system was not the expected Lockheed-developed RSI silica tiles. Instead, McDonnell Douglas proposed using mullite HCF shingles[20] that were glued to the bottom and side of the orbiter over a thin layer of foam for strain relief. The top of the orbiter would use ESA-3560 or SLA-561 ablator, as would the leading edges of the wing and vertical stabilizer.[21]

Like Grumman, the engineers at North American Rockwell projected a 500-mission life for the orbiter. To accomplish this, the company selected a thermal protection system designed to subject the primary structure to no more than 350 °F. North American was also not completely convinced that the Lockheed-developed silica tiles would mature quickly enough for the early orbital flights, so the baseline used mullite HCF shingles instead. North American continued to monitor the progress being made on the silica tiles since they were lighter weight, and the contractor expected to switch to them at an appropriate time. The leading edges and nose cap were carbon-carbon, and no substantial use of ablators was made anywhere on the orbiter.[22]

Orbiter Contract Award

James Fletcher, George Low, and Richard McCurdy met the morning of July 26, 1972, for the final review of the Space Shuttle orbiter proposals.[23] They noted that the evaluation team had ranked McDonnell Douglas and Lockheed significantly below the other two competitors, so they concentrated instead on North American and Grumman. After carefully reviewing the mission suitabil-

19. Proposal to NASA-MSC: Space Shuttle (multiple volumes), Grumman report LMSC-D157364 (Sunnyvale, CA: Lockheed Missiles and Space Company, May 12, 1972).

20. For some reason, the mullite material was composed of shingles, and the Lockheed RSI used tiles.

21. *Space Shuttle Program* (multiple volumes), McDonnell Douglas report MDC-E0600 (St. Louis, MO: McDonnell Douglas, May 12, 1972).

22. Proposal for the Space Shuttle Program (multiple volumes), North American report SD72-SH-50-3 (Downey, CA: North American Rockwell Space Division, May 12, 1972).

23. At this time, Fletcher was Administrator, Low was Deputy Administrator, and McCurdy was Associate Administrator for Organization and Management.

ity scores, the three men decided that the advantage went to North American. Since North American also presented the lowest probable cost, it was deemed the winner of the contract.[24]

After the stock market closed on July 26, 1972, NASA announced that the Space Transportation Systems Division of North American Rockwell[25] had been awarded the $2.6 billion contract to design and build the Space Shuttle orbiter.[26] The design was loosely based on the final MSC-040C configuration, and during its return to Earth, the vehicle would be capable of a 1,265-mile reentry maneuver on either side of its flightpath, thereby meeting Air Force requirements.[27]

The Space Shuttle represented the largest hypersonic vehicle ever designed, providing the first real test of experimental and theoretical knowledge gained during the X-15, ASSET, and PRIME programs. No real precedents existed to help establish the design requirements for such a vehicle, something the flight-testing of Dyna-Soar would have significantly helped. A primary challenge for the designers was the preflight prediction of the aerodynamic characteristics of the vehicle with sufficient accuracy to conduct the first orbital flight with a crew. This also required overcoming the unknowns involved in hypersonic wind tunnel testing, and developing reliable simulation techniques without having an extensive database to compare the results.[28]

In its Phase C response, Rockwell proposed using mullite shingles manufactured from aluminum silicate instead of the Lockheed-developed silica tiles since the technology was better understood and more mature. But the mullite tiles were heavier and potentially less durable. Given the progress Lockheed was making, NASA and Rockwell asked the Battelle Memorial Institute to evaluate both candidates' systems—an evaluation the Lockheed product won. Nevertheless, the Lockheed material was not appropriate for the very high-temperature areas of the orbiter, such as the nose cap and wing leading edge. These

---

24. Memorandum for Record, James C. Fletcher, George M. Low, and Richard McCurdy, September 18, 1972, in the files of the NASA History Division, conveniently reprinted in *Exploring the Unknown: Accessing Space*, vol. IV (Washington, DC: NASA SP-4407, 1999), pp. 262–268.

25. North American Rockwell became Rockwell International on February 16, 1973.

26. *Exploring the Unknown*, vol. IV, quotes on pp. 266–267.

27. NASA News Release, July 26, 1972; Major Thomas W. Rutten, "History of the Directorate of Space DCS/Research and Development For the Period 1 July 1972 to 31 December 1972" (Los Angeles, CA: U.S. Air Force, January 1973), in the files of the AFHRA K140-01.

28. J.C. Young, J.M. Underwood, E.R. Hillje, A.M. Whitnah, P.O. Romere, J.D. Gamble, B.B. Roberts, G.M. Ware, W.I. Scallion, B. Spencer, Jr., "The Aerodynamic Challenges of the Design and Development of the Space Shuttle Orbiter," NASA, 1985.

would use a reinforced carbon-carbon (RCC) material originally developed by Ling-Temco-Vought (LTV), based on research for the Dyna-Soar program.[29]

Interestingly, NASA and Rockwell originally believed that the leeward side (top) of the orbiter would not require any thermal protection. However, in March 1975, the AFFDL conducted a briefing for Space Shuttle engineers on the classified results of the ASSET, PRIME, and boost-glide reentry vehicle programs, which indicated leeward-side heating was a serious consideration. The thermal environment was not particularly severe but easily exceeded the 350 °F capability of the aluminum skin. To alleviate this concern, Rockwell would cover the upper surfaces of the wings and aft fuselage, along with the entire payload bay door (areas that never exceeded 650 °F) with an ablative elastomeric reusable surface insulation (ERSI). The ERSI, which consisted of a silicone resin with titanium dioxide and carbon-black pigments for thermal control, would be bonded directly to the aluminum airframe.

Between April 1978 and January 1979, a team from the AFFDL conducted a review of potential orbiter heating concerns and concluded that the OMS pod might have unanticipated problems. Air Force researchers conducted a series of tests at the Arnold Engineering Development Center between May and November 1979, and further tests were run at the Naval Surface Warfare Center shock tunnel in May 1980. The researchers discovered that the composite OMS pod skin would deflect considerably more than expected. The thin 8-by-8-inch RSI tiles were relatively weak under bending loads, and engineers feared that the tiles might fracture and separate from the vehicle. Since the tiles had already been installed, a unique solution was developed in which each tile was diced while still attached to the pod by carefully cutting it into nine equal parts. The technique proved so successful that it has subsequently been used on other areas of the orbiter.[30]

The proposed thermal protection system at this point differed somewhat from the system ultimately used on the production orbiters. ERSI ablator covered the upper surfaces of the wings and payload bay doors, where temperatures were always less than 650 °F. Black LI-2200 high-temperature reusable surface insulation (HRSI) tiles protected the bottom of the fuselage and

---

29. Paul A. Cooper and Paul F. Holloway, "The Shuttle Tile Story," *Astronautics & Aeronautics,* January 1981, pp. 24–34; "Technology Influence on Space Shuttle Development," Eagle Engineering Report 86-125C (June 8, 1986), pp. 6-4 and 6-5; W. M. Pless, "Space Shuttle Structural Integrity and Assessment Study, External Thermal Protection System," Lockheed report CR-134452/LG73-ER0082 (Marietta, GA: Lockheed-Georgia Company, June 1973).

30. Cooper and Holloway, "Shuttle Tile Story," pp. 24–27; Schneider and Miller, "Scales of the Bird," pp. 409–410; Hallion and Young, "Fulfillment of a Dream," pp. 1160–1163.

This 1979 structural test article of the Space Shuttle wing shows the reinforced carbon-carbon and tile structure. NASA.

wings, the entire vertical stabilizer, and most of the forward fuselage where temperatures exceeded 650 °F but were less than 2,300 °F. The tile thickness varied as necessary to limit the aluminum airframe's maximum temperature to 350 °F. Actually, earlier studies showed a combination of LI-2200 and LI-3000 tiles would be required, although this was later reduced to entirely LI-2200, then to a combination of LI-900 and LI-2200.[31] Further study showed some drawbacks to this protection system, and when Columbia was manufactured the ERSI would be replaced by a combination of LI-900 tiles and Nomex felt blankets. The HRSI tiles were used over a smaller percentage of the orbiter than originally anticipated as further analysis showed that entry heating would not be as severe as expected, mainly because of improved flight profiles.[32]

In the meantime, a problem had developed with the tiles themselves. As flight profiles were refined and aeroloads better understood, engineers began to question whether the tiles could survive the harsh conditions. By mid-1979 it had become obvious that the tiles in certain areas did not have sufficient strength to survive the tensile loads of a single mission. NASA immediately began an extensive search for a solution that eventually involved outside blue-ribbon panels, Government agencies, academia, and most of the aerospace industry. As Space Shuttle Program Deputy Director LeRoy Day recalled, "There is a case [the tile crisis] where not enough engineering work, probably, was done early enough in the program to understand the detail–the mechanical properties–of this strange material that we were using. It was, potentially, a showstopper."[33]

As engineers spent more time examining the problem, they realized that the issue was not with the individual tiles but the manner in which the tiles were attached to the underlying structure. Analysis indicated that while each individual component—the tile, the felt strain-isolation pads (SIP) under the tile, and the two layers of adhesives—had satisfactory tensile strength, the components when combined as a system lost about 50 percent of their strength. Engineers largely attributed this to stiff spots in the SIP (caused by needling) that allowed the system strength to decline to as low as 6 pounds per square inch, instead of the baseline 13 pounds per square inch.[34]

---

31. Pless, "Space Shuttle Structural Integrity and Assessment Study."

32. Ibid.

33. First quote from Cooper and Holloway, "Shuttle Tile Story," p. 25. Day quote from an interview of LeRoy E. Day by John Mauer, October 17, 1983, pp. 5–6, in the files of the JSC History Office, Houston, TX.

34. William C. Schneider and Glenn J. Miller, "The Challenging 'Scales of the Bird'" (Shuttle Tile Integrity), paper presented at the Space Shuttle Technical Conference (CR-2342) at JSC, June 28–30, 1983, pp. 403–413.

The Space Shuttle structural test article undergoing tests in the Lockheed facility in Palmdale, CA. This airframe, STA-099, was subsequently refitted for orbital flight and became the Challenger (OV-099). NASA.

In October 1979, engineers developed a densification process that involved filling voids between fibers at the inner mold line (the part next to the SIP pad) with a special slurry mixture consisting of Ludox (a colloidal silica made by DuPont) and a mixture of silica and water. Since the tiles had been water-proofed during manufacture, the process began by applying isopropyl alcohol to dissolve the waterproof coating, then painting the back of the tile with Ludox. After air-drying for 24 hours, the tiles were baked in an oven at 150 °F for 2 hours, then re-waterproofed using Dow Coming's Z-6070 methyltrime-thoxysilane.[35] The densified layer acted as a plate on the bottom of the tile, eliminating the effect of the local stiff spots in the SIP, bringing the total system strength back up to 13 pounds per square inch.[36]

Installing the tiles on the vehicle presented its own problems, and Rockwell quickly ran out of time to do so while Columbia was in Palmdale, CA. NASA

---

35. Probably the most famous silane product is 3M Scotchgard™.
36. Schneider and Miller, "The Challenging 'Scales of the Bird'," pp. 403–413.

needed to present the appearance of maintaining a schedule, and Columbia moving to KSC was a very visible milestone. So in March 1979, Columbia was flown from Palmdale to KSC on the 747 Shuttle Carrier Aircraft and was quickly moved into the Orbiter Processing Facility. Just over 24,000 tiles had been installed in Palmdale, with 6,000 left to go. Unfortunately, by now it appeared that all of the tiles would need to be removed so that they could be densified.

The challenge revolved around how best to salvage as many of the installed tiles as possible while ensuring sufficient structural margin for a safe flight. The approach that engineers developed to overcome this almost insurmountable challenge was called the tile proof test. Technicians induced a stress over the entire footprint of each tile equal to 125 percent of the maximum flight stress experienced at the most critical point on the tile footprint.[37]

The device developed for the proof test used a vacuum chuck to attach to the tile, a pneumatic cylinder to apply the load, and six pads attached to surrounding tiles to react the load. Since any appreciable tile load might cause internal fibers to break, acoustic sensors placed in contact with the tiles were used to monitor the acoustic emissions for any internal fiber breakage. This testing proved the majority of the tiles were adequately bonded for flight, and only 13 percent failed the proof test; those were replaced with densified tiles.[38]

Engineers also developed two other techniques to strengthen tiles while they were still on the vehicle. The first involved relatively small, thick tiles that were usually located on the underside of the orbiter. As shock waves swept air over the tiles, they tended to rotate, inducing high stresses at the SIP bond. Technicians installed gap fillers that prevented the tiles from rotating, but this solution was not effective for small thin tiles, so engineers developed a technique in which the filler bar surrounding the SIP was bonded to the tiles. Technicians inserted a crooked needle into the tile-to-tile gap and posited RTV adhesive on top of the filler bar. This significantly increased the bonded footprint and decreased the effects of a shock-imposed overturning moment.

For the next 20 months, technicians at KSC worked 3 shifts per day, 7 days per week testing and installing 30,759 tiles. By the time the tiles were installed, proof tested, often removed and reinstalled, then re-proof tested, the technicians averaged 1.3 tiles per person per week. In June 1979, Rockwell estimated that 10,500 tiles needed to be replaced; by January 1980, over 9,000 of these had been installed, but the number remaining had ballooned to 13,100 as additional tiles failed the proof test or were otherwise damaged. By September

---

37. Ibid.
38. Ibid.

During the flight of STS-1 in April 1981, Columbia shed some of its thermal protection system tiles, as shown in this orbital imagery. The vehicle returned safely. NASA.

1980, only 4,741 tiles remained to be installed, and by Thanksgiving, the number was below 1,000. It finally appeared that the end was in sight.[39]

Thermal Protection System

The thermal protection system protected the aluminum structure of the orbiter during the atmospheric portion of flight (both ascent and entry), maintaining skin temperatures at less than 350 °F. The chosen thermal protection system was completely passive in operation and, in theory, capable of unlimited reuse with only minor refurbishment. It provided a smooth aerodynamic mold line using lightweight materials with low thermal conductivity.[40] These materials had to perform in temperature ranges from −250 °F in the cold soak of space to entry temperatures approaching 3,000 °F. The peak heating rates and

---

39. Richard P. Hallion and James O. Young, "Space Shuttle: Fulfillment of a Dream," Case VIII of *The Hypersonic Revolution: Case Studies in the History of Hypersonic Technology, Volume II, From Scramjet to the National Aero-Space Plane (1964–1986)* (Washington, DC: USAF Histories and Museums Program, 1998), p. 1166.

40. M. Fleming et al., "A History of TPS Failures on Space Shuttle Orbiter," Rockwell IL-284-400-94-139 (October 15, 1994), a report provided to the Columbia Accident Investigation Board by Boeing.

longest exposures occurred during high cross-range entry maneuvers when equilibrium surface temperatures range from 3,000 °F, at stagnation points on the nose and leading edges of the wing and tail down to 600 °F on leeward surfaces of the wing.[41]

The thermal protection system was composed of RCC, four types of RSI tiles, two types of RSI blankets, thermal barriers, gap fillers, thermal window-panes, and thermal seals.[42]

Reinforced Carbon-Carbon

Where temperatures exceeded 2,300 °F, such as the wing leading edge and the nose cap, the vehicle featured a composite composed of pyrolyzed carbon fibers in a pyrolyzed carbon matrix with a silicon carbide coating. RCC weighs 90 to 100 pounds per cubic foot. The operating range of RCC is from −250 °F to 3,000 °F. The carbon-carbon nose cap and wing leading edges were manufactured by LTV (now Lockheed Martin) of Grand Prairie, TX.[43] The wing leading edge consisted of 22 RCC panels, and because of the wing profile changes from inboard to outboard, each RCC panel was unique.[44] It should be noted that RCC was not an insulator; it was simply a structure that could withstand the appropriate aero and thermal loads. The underlying structure needed to be protected from extreme heat by other means.

RCC fabrication begins with a graphite-saturated rayon cloth impregnated with a phenolic resin. This impregnated cloth is layed up as a laminate and cured in an autoclave. After curing, the laminate is pyrolyzed to convert the resin to carbon. This is then impregnated with furfural alcohol in a vacuum chamber, then cured and pyrolyzed again to convert the furfural alcohol to carbon. This process is repeated three times until the desired carbon-carbon properties are achieved.[45]

---

41. *Shuttle Crew Operations Manual*, p. 1.2-14.

42. *Shuttle Crew Operations Manual*, p. 1.2-15; STS-59 Press Kit, April 1994, p. 37; *Ames Astrogram* (newsletter), April 1, 1994; Charles Redmond, "Improved Shuttle Tile to Fly on STS-59," NASA News Release 94-54, March 31, 1994; "Shuttle Thermal Protection," LTV News Release V84-29, February 1, 1990; Chin Panel, LTV News Release, March 23, 1988; various material data sheets on TPS products, Lockheed Missiles and Space Company, January 1984.

43. M. Fleming et al., "A History of TPS Failures."

44. *Columbia Accident Investigation Board Final Report* (Washington, DC, NASA, August 2003), pp. 55–56.

45. Donald M. Curry, John W. Latchen, and Garland B. Whisenhunt, "Space Shuttle Leading Edge Structural Development," AIAA paper 83-0483, presented at the 21st Aerospace Sciences Meeting, January 10–13, 1983; *Shuttle Crew Operations Manual*, p. 1.2-14.

To provide sufficient oxidation resistance to enable reuse, the outer layers of the RCC are converted to silicon carbide. The RCC is packed in a retort with a dry-pack material made of a mixture of alumina, silicon, and silicon carbide. The retort is placed in an argon-purged furnace with a stepped-time-temperature cycle up to 3,200 °F. A diffusion reaction occurs between the dry pack and carbon-carbon that converts the outer layers of the carbon-carbon to silicon carbide (whitish-gray color). The part is then sprayed with tetraethyl orthosilicate and sealed with a glossy overcoat. The RCC laminate is superior to a sandwich design because it is lightweight and rugged; it also promotes internal cross-radiation from the hot stagnation region to cooler areas, thus reducing stagnation temperatures and thermal gradients around the leading edge.

The RCC panels were mechanically attached to the wing leading edge with a series of floating joints to reduce loading on the panels caused by wing deflections. A T-seal between each panel allowed lateral motion and thermal-expansion differences between the RCC and the orbiter wing. The T-seals were also constructed of RCC.

Since RCC was highly radiative and provided no thermal protection, the adjacent aluminum structure had to be protected by internal insulation. The forward wing spar and nose bulkhead were protected by a series of captive and non-captive RSI tiles that were installed onto removable carrier panels with Inconel 601–covered Dyna-Flex (Cerachrome) bulk insulation (also called Incoflex).[46]

Beginning in 1998, engineers began modifying the leading-edge RCC to enable it to withstand larger punctures. The original design allowed a 1-inch hole in the upper surface of any panel. But on the lower surface, no penetrations were allowed on panels 5 through 13 because any hole generated by orbital debris (or other causes) would allow heat from the plasma flow during entry to quickly erode the 0.004-inch Inconel foil on the Incoflex insulators, exposing the leading-edge attach fittings and wing-front spar to the direct blast of hot plasma. The upgrade included additional insulation that was able to withstand a penetration of up to 0.25 inch in diameter in the lower surface of panels 9 through 12, and up to 1 inch on panels 5 through 8 and 13. To achieve this, Nextel 440 fabric insulation was wrapped around the Incoflex insulators—one layer for panels 5 through 7 and 11 through 13, and two fabric layers for panels 8 through 10 (which have the highest-potential heating environment).[47]

Despite the modification, an ascent debris strike on an RCC panel on the left wing of Columbia during STS-107 resulted in the destruction of the orbiter

---

46. Fleming et al., "A History of TPS Failures on Space Shuttle Orbiter."

47. "Space Shuttle Orbiter Mass Properties Status Report"; *Space Shuttle Program 1999 Annual Report*, part 2, pp. 25–26.

There were difficulties with burn-throughs on the reinforced carbon-carbon leading edge, as shown in this comparison of STS-2 and STS-3 panels in 1982. NASA.

during entry. This led to a reevaluation of the relative strength of the material, and subsequent testing revealed that it was not as robust as once thought, particularly to direct debris strikes. Nevertheless, it was one of the few materials that can withstand the environment it is subjected to, and it continued to be used, although with somewhat more inspection and analysis.

Reusable Surface Insulation Tiles

The RSI tiles used on the orbiter had three different densities of the base silica material. LI-900 weighed 9 pounds per cubic foot and comprises 85 percent of the orbiter's tiles. Denser LI-2200 tiles weighed 22 pounds per cubic foot and were used around door and access panels, making up approximately 1 percent of the orbiter's tiles. FRCI-12 tiles weighed 12.5 pounds per cubic foot and made up approximately 12 percent of the orbiter's tiles. The secret to the tiles' performance was the fact that they were 80 to 90 percent void (contain air) and minimized heat transfer within the tile itself.[48]

---

48. Fleming et al., "A History of TPS Failures on Space Shuttle Orbiter."

Black HRSI tiles were used on the entire underside of the vehicle where RCC was not used, including the following surfaces: the base heat shield; selected areas on the upper forward fuselage, such as around the forward fuselage windows; the portions of the OMS/RCS pods; the leading and trailing edges of the vertical stabilizer; the wing-glove areas; the elevon trailing edges; the area adjacent to the RCC on the upper wing; the interface with wing leading edge RCC; and the upper body flap. The HRSI tiles protected areas where temperatures were between 1,200 °F and 2,300 °F.

The HRSI tiles were made of low-density, high-purity, 99.8-percent amorphous silica fibers derived from common sand; they were 1 to 2 millimeters thick and were made rigid by ceramic bonding. A slurry containing fibers mixed with water was frame-cast to form soft, porous blocks onto which a colloidal silica binder solution was added. When it was sintered, a rigid block was produced that was cut into quarters and then machined to the precise dimensions required for individual tiles. HRSI tiles varied in thickness from 1 to 5 inches, with the thickness determined by the heat load encountered during entry. Generally, the HRSI tiles were thicker at the forward areas of the orbiter and thinner toward the aft end. The HRSI tiles were nominally 6-by-6-inch squares, but they varied in size and shape in the closeout areas. The HRSI tiles withstood on-orbit cold soak conditions, repeated heating and cooling thermal shock, and extreme acoustic environments (163 decibels) at launch.[49]

The HRSI tiles were coated on the top and sides with a mixture of powdered tetrasilicide and sintered borosilicate reaction-cured glass (RCG) coating with a liquid carrier. The tiles were individually sprayed with 10 to 15 coats of the coating slurry to produce a final coating weight of 0.09 to 0.170 pounds per square foot. The coated tiles then were placed in an oven and heated to 2,300 °F for 90 minutes. This resulted in a black, glossy coating that had a surface emittance of 0.85 and a solar absorptance of about 0.85. Then, the tiles were waterproofed using heated methyltrimethoxysilane. The silane vapor penetrated the tile, rendering the fibrous silica material hydrophobic.[50]

As an aside, an HRSI tile taken from a 2,300 °F oven could be immersed in cold water without damage. Also, surface heat dissipated so quickly that an uncoated tile could be held by its edges with an ungloved hand only seconds after removal from the oven while its interior still glowed red.

As the higher density HRSI tiles were replaced, new fibrous refractory composite insulation (FRCI) tiles were usually used instead. Researchers at NASA Ames developed this product later in the program, and it was more durable

---

49. Ibid.
50. Ibid.

When an orbiter returned from a mission, technicians carefully inspected and replaced thermal protection system tiles as needed. Although the tiles required significant maintenance early in the program, after a few years they became very reliable, and late in the program only a handful of tiles were replaced after a mission. NASA.

and resistant to cracking than the earlier HRSI. The tiles were essentially similar to the original HRSI tiles, with the addition of a 3M Company additive called Nextal (AB312—alumina-borosilicate fiber). With an expansion coefficient 10 times that of the 99.8-percent pure silica, Nextel activated a boron fusion, which effectively welded the pure silica fibers into a rigid structure during sintering. The resulting composite-fiber refractory material, composed of 20 percent Nextal and 80-percent pure silica, had entirely different physical properties than the original 99.8-percent pure silica tiles.

The RCG (black) coating of the FRCI-12 tiles was compressed as it was cured to reduce the coating's sensitivity to cracking during handling and operations. In addition to the improved coating, the FRCI-12 tiles were about 10 percent lighter than the HRSI tiles for a given thermal protection. FRCI-12 tiles have demonstrated a tensile strength at least three times greater than that of the HRSI tiles and a use temperature approximately 100 °F higher.

White low-temperature reusable surface insulation (LRSI) tiles were used in selected areas of the forward fuselage, midfuselage, aft fuselage, vertical stabilizer, upper wing, and OMS/RCS pods. These tiles protected areas where temperatures were between 750 °F and 1,200 °F, and they had a white surface coating to provide better thermal characteristics while on orbit.

LRSI tiles were of the same construction and had the same basic character-istics as the HRSI tiles but were cast thinner (0.2 to 1.4 inches). The tiles were produced in 8-by-8-inch squares and had a white RCG optical and moisture-resistant coating applied 10 millimeters thick to the top and sides. The coating was made of silica compounds with shiny aluminum oxide to obtain optical properties, and it provided on-orbit thermal control for the orbiter. The LRSI tiles were treated with bulk waterproofing and were installed on the orbiter in the same manner as the HRSI tiles. The LRSI tile had a surface emittance of 0.80 and a solar absorptance of 0.32.

A new high-temperature tile, known as toughened unipiece fibrous insula-tion (TUFI), was developed by NASA Ames in 1993 and was first flown on STS-59. The results were encouraging enough for TUFI tiles to be installed on an attrition basis to replace 304 HRSI tiles on the base heat shield around the Space Shuttle main engine and lower body-flap surface. TUFI was the first of a new type of composite known as functional gradient materials in which the density of the material is relatively high on the outer surface and becomes increasingly lower within the insulation. Unlike the HRSI tiles, in which the tetrasilicide and borosilicate glass coating received little support from the underlying tile, TUFI's outer surface was fully integrated into the insula-tion, resulting in a more damage-resistant tile. TUFI tends to dent instead of shatter when hit. This tile material is also known as an alumina-enhanced thermal barrier (AETB-8), or TUFI/RCG-coated tiles.

Regardless of type, every tile was unique to a location on the orbiter, cut to size, thickness, and shape to conform to the contours of the vehicle and the necessary insulative properties. Using a commercial high-temperature paint called Spearex, the tiles were then permanently marked with a part number that designated its location on the orbiter (X/Y coordinate) and with a unique serial number that provided traceability by lot number to the materials and processes affecting the individual tile. Also, the part-number identification was located on the forward or leading edge of the tile to assist in the correct orientation of the tile at the time of installation. Because of discontinuities in the thin RCG coating, caused either by processing or flight damage, the tiles required re-waterproofing after each flight. Each tile was pneumatically injected with dimethylethoxysilane, which chemically reacted with the silica fibers of the base material and prevented water absorption into the tile.[51]

Since the tiles could not withstand airframe-load deformation, stress isola-tion was necessary between the tiles and the orbiter structure. This was provided by SIPs made of Nomex felt in thicknesses of 0.090, 0.115, or 0.160 inch.

---

51. Ibid.

The SIP was bonded to the tiles, and then bonded to the orbiter skin using a silicon RTV adhesive. However, the SIP introduced stress concentrations at the needled fiber bundles. This resulted in localized failure in the tile just above the RTV bond line. To solve this problem, the inner surface of the tile was densified to distribute the load more uniformly. The densification process used a Ludox ammonia-stabilized binder. When mixed with silica slip particles it became a cement, and when mixed with water it dried to a finished hard surface. A silica-tetraboride coloring agent was mixed with the compound for penetration identification. The densification coating penetrated the tile to a depth of 0.125 inch, and the strength and stiffness of the tile and SIP system were increased by a factor of two.

The tiles thermally expanded or contracted very little compared to the orbiter structure, so it was necessary to leave 0.025- to 0.065-inch gaps between them to prevent tile-to-tile contact. Nomex felt-material insulation was required in the bottom of the gap between tiles. These filler bars were supplied in thicknesses corresponding to the SIPs, cut into strips 0.75-inch wide, and bonded to the orbiter structure. The filler bar was waterproof and temperature-resistant up to approximately 800 °F.

White blankets made of coated Nomex felt reusable surface insulation (FRSI) were used on the upper payload bay doors, portions of the midfuselage and aft-fuselage sides, portions of the upper wing surface, and parts of the OMS/RCS pods. The FRSI blankets protected areas where temperatures were below 700 °F. The FRSI contained the same Nomex material as the SIP used under the tiles. The FRSI blankets generally consisted of 3-by-4-foot, 0.16-to-0.40-inch-thick sheets, depending on location. A white-pigmented 92-007 silicon elastomer coating functioned as aerodynamic-erosion protection and as a moisture barrier, and it acted as a passive cooling system by having low solar absorptance and high reflectance of solar radiation. The FRSI had an emittance of 0.8 and solar absorptance of 0.32 and covered nearly half of the orbiter's upper surfaces. The density of the coated material was 5.4 pounds per cubic foot.[52]

Nomex felt is a noncombustible, heat-resistant aromatic polyamide (aramid) fiber that is 2 deniers in fineness, 3 inches long, and crimped. These were loaded into a carding machine that untangled the clumps of fibers and combed them to make a tenuous mass of lengthwise-oriented, relatively parallel fibers called a web. The cross-lapped web was fed into a loom, where it was lightly needled into a batt. Generally, two such batts were placed face-to-face and needled together to form felt. The felt then was subjected to a multi-needle pass process

---

52. Ibid.

that passed barbed needles through the fiber web in a sewinglike procedure that compacted the transversely oriented fibers into a pad. This needle-punch operation was repeated as many times as required to produce a felt product with specified physical properties. The needled felt was calendered to the proper thickness by passing through heated rollers at selected pressures. The calendered material was heat-set at 500 °F to thermally stabilize the felt. The FRSI was bonded directly to the orbiter by RTV silicon adhesive, 0.20 inch thick.[53]

After the initial delivery of Columbia, an advanced flexible reusable surface insulation (AFRSI) was developed as a lightweight replacement for the thin, fragile LRSI tiles. The AFRSI blankets protected areas where temperatures were below 1,200 °F and aerodynamic loads were minimal. AFRSI was a quilted blanket consisting of a quartz fibrous batt insulation and a woven quartz fiber outer fabric quilted on 1-inch centers to a glass fiber inner fabric. The quilting was accomplished using a Teflon-coated quartz thread with three to four stitches per inch. The sewn quilted fabric blanket was manufactured in 3-by-3-foot squares with an overall thickness of 0.25 to 2 inches. The material had a density of 8 to 9 pounds per cubic foot and was used where surface temperatures ranged from 750 °F to 1,200 °F. The blankets were cut to the planform shape required and bonded directly to the orbiter by RTV silicon adhesive 0.20 inch thick.[54]

The first application of AFRSI was on the OMS pods of Columbia during STS-6. AFRSI blankets were used on Discovery and Atlantis to replace the majority of the LRSI tiles. The LRSI tiles on Columbia's midfuselage, payload bay doors, and vertical stabilizer were replaced with AFRSI blankets during the Challenger standdown. Endeavour was delivered with an even greater use of AFRSI, and the other orbiters (except OV-102) initially migrated toward this configuration during normal maintenance as well as major overhauls. These blankets were initially used on the fuselage, upper elevens, upper wings, vertical stabilizer, and forward canopy surfaces. However, in an effort to save as much weight as possible for missions to the International Space Station (ISS), much of the AFRSI on the mid fuselage and aft fuselage, payload bay doors, and upper wing surfaces of the three ISS orbiters was subsequently replaced by the lighter FRSI during Orbiter Maintenance Down Period (OMDP). During Endeavour's first OMDP, 1,472 pounds were saved by this modification. A change to the flipper doors that included FRSI saved another 520 pounds.[55]

---

53. Ibid.

54. Ibid.; *Shuttle Crew Operations Manual*, p. 1.2-15.

55. STS-89 Orbiter Rollout Milestone Review.

As of its last flight, in 2011, Atlantis[56] was protected with 18,490 HRSI-9 (LI-900) tiles, 2,846 FRCI-12 tiles, 469 AETB-8 (TUFI) tiles, 234 Boeing Replacement Insulation (BRI)-18 tiles, and 3,254 thermal blankets of various types. Approximately 20 tiles were typically damaged on each flight, and the average postflight refurbishment consisted of about 200 tiles and 10 blankets. However, approximately half of those were replaced for reasons unrelated to their failure. These reasons included removing a tile (or blanket) to gain access to the structure or component it covers for maintenance, or replacing the tile (or blanket) with an improved version to solve some perceived (or anticipated) anomaly.

Thermal barriers prevented hot air and plasma from flowing through the various penetrations in the orbiter, such as the landing-gear doors and crew hatch. The thermal barriers were constructed of tubular Inconel 750 wire mesh filled with alumina silica fibrous insulation, or Saffil, wrapped into a ceramic sleeving and a Nextel AB312 ceramic alumina-borosilicate fabric outer cover. A black ceramic emittance coating was applied to the outer surface and an RTV silicone rubber was used to attach the barrier to the structure. Thermal barriers were designed for service in areas seeing temperatures up to 2,000 °F.[57]

Because the tiles were installed with a space between them, it was sometimes necessary to use gap fillers to ensure that heat did not seep between the tiles and impact the aluminum skin. Two types of gap fillers were used: the pillow/pad and the Ames type. The pillow/pad gap filler was made of a ceramic fabric with alumina silica insulation with two layers of 0.001-inch-thick Inconel foil inside for support. The pad gap filler was stitched with quartz thread, and a black ceramic emittance coating was applied to the outer mold line. A coating of red RTV-560 was applied to the bottom surface as a stiffener to aid with installation into the tile-to-tile gaps. Approximately 5,000 pad-type gap fillers, each 0.200 inch thick, were used on the lower body flap, lower elevons, and vertical stabilizer. The Ames-type gap filler was made from a ceramic fabric that was originally impregnated with a black silicone coating, although this was subsequently replaced with a more durable ceramic coating.[58]

Two types of internal insulation blankets were used: fibrous bulk and multi-layer. The bulk blankets were fibrous materials with a density of 2 pounds per cubic foot and a sewn cover of reinforced-Kapton acrylic film. The cover material had 13,500 holes per square foot for venting. Acrylic-film tape was used for cutouts,

---

56. This varied slightly for each vehicle, with Columbia having additional HRSI-22 tiles on the vertical stabilizer (to protect the SILTS pod) and additional HRSI-9 tiles on the upper wing leading edge chines, and Endeavour having fewer tiles due to the extensive use of AFRSI.

57. Fleming et al., "A History of TPS Failures on Space Shuttle Orbiter."

58. Ibid.

patching, and reinforcements. Tufts throughout the blankets minimized billow-
ing during venting. The multilayer blankets were constructed of alternate layers of
perforated-Kapton acrylic-film reflectors and Dacron-net separators. There were 16
reflector layers in all, with the two-cover halves counting as two layers. The covers,
tufting, and acrylic-film tape were similar to that used for the bulk blankets.[59]

Flags and letters were painted on the orbiter with a Dow Corning-3140
silicon-based material, colored with pigments. It was basically the same paint
used to paint automobile engines, and it would break down in temperature
ranges between 800 °F and 1,000 °F. Because of this, almost all markings were
painted in relatively low-temperature areas of the orbiter.

Space Shuttle Entry

After completing its on-orbit tasks, the Space Shuttle crew oriented the orbiter
to a tail-first attitude using the RCS, then used the two OMS engines to slow
the orbiter for deorbit. The RCS turned the orbiter's nose forward for an entry
that began at 400,000 feet (defined as the entry interface), slightly over 5,000
miles from the landing site. The orbiter's velocity at entry was approximately
17,500 mph with a 40-degree angle of attack. The forward RCS thrusters were
inhibited by the onboard computers immediately before entry, and the aft RCS
thrusters maneuvered the vehicle until a dynamic pressure of 10 pounds per
square feet was sensed, which was when the orbiter's ailerons became effective.
The aft RCS roll thrusters were then deactivated. At a dynamic pressure of 20
pounds per square foot, the orbiter's elevators became effective and the aft
RCS pitch thrusters were deactivated. The orbiter's speed brake was used below
Mach 10.0 to induce a more positive downward elevator-trim deflection. At
approximately Mach 3.5, the aerodynamic rudder was activated, and the aft
RCS yaw thrusters were deactivated at 45,000 feet.[60]

Entry guidance had to dissipate the tremendous amount of energy the
orbiter possessed when it entered the atmosphere to ensure that the vehicle did
not either burn up (entry angle too steep) or skip back out of the atmosphere
(entry angle too shallow). During entry, excess energy was dissipated by the
atmospheric drag on the orbiter. A steep trajectory gave higher atmospheric-
drag levels, which resulted in faster energy dissipation. Normally, the angle of
attack and angle of roll enabled the atmospheric drag of any flight vehicle to

---

59. *Shuttle Crew Operations Manual*, p. 1.2-14.

60. "Operations," *National Space Transportation System Reference*, vol. 2 (Washington, DC: NASA,
   September 1988), pp. 75–90.

be controlled. However, for the orbiter, angle-of-attack variation was rejected because it created exterior surface temperatures in excess of the capabilities of the thermal protection system. The angle-of-attack schedule was loaded into the computers as a function of relative velocity, leaving only roll angle for energy control. Increasing the roll angle decreased the vertical component of lift, causing a higher sink rate and increased energy dissipation. Increasing the roll angle raised the surface temperature of the orbiter somewhat, but not nearly as drastically as an equivalent angle-of-attack variation.

If the orbiter was low on energy (current range-to-go greater than nominal), entry guidance would command lower-than-nominal drag levels. If the orbiter had too much energy (current range-to-go less than nominal), entry guidance would command higher-than-nominal levels to dissipate the extra energy to within the orbiter's limits to withstand the additional surface heating. The goal was to maintain a constant heating rate until the orbiter was below 13,000 mph. During entry, the temperatures on the wing leading edge increased steadily for about 10 minutes, at which point the orbiter was traveling at Mach 23 and at 230,000 feet with wing leading-edge temperatures of about 3,000 °F. While the orbiter was descending, superheated air molecules discharged light that could be seen as bright flashes through the crew cabin windows.[61]

Roll angle was also used to control cross range. Azimuth error is (1) the angle between the plane containing the orbiter's position vector and heading-alignment cylinder tangency point and (2) the plane containing the orbiter's position vector and velocity vector. When the azimuth error exceeded a pre-determined number, the orbiter's roll angle was reversed.

The equilibrium-glide phase shifted the orbiter from the rapidly increasing drag levels of the temperature-control phase to the constant-drag phase. The equilibrium-glide flight was defined as flight in which the flightpath angle—the angle between the local horizontal and the local velocity vector—remained constant. Equilibrium-glide flight provided the maximum downrange capability and lasted until the drag acceleration reached 33 ft/sec squared. At that point, the constant-drag phase began. The angle of attack was initially 40 degrees but ramped down toward 36 degrees at the end of the phase. In the transition phase, the angle of attack continued to ramp down, reaching approximately 14 degrees as the orbiter reached the terminal area energy management (TAEM) interface at approximately 83,000 feet and Mach 2.5.

The TAEM interface steered the orbiter to the nearest of two heading-alignment circles (HAC), whose radii were approximately 18,000 feet and which were located tangent to and on either side of the runway centerline

---

61. *Columbia Accident Investigation Board Final Report* (Washington, DC: NASA, August 2003), pp. 38–40.

STS-4 sustained some damage to the thermal protection system, but the orbiter returned to Earth safely. NASA S82-28874.

on the approach end. In TAEM guidance, excess energy was dissipated with S-turns, and the speed brake was used to modify the drag coefficient as required. This decreased the ground-track range as the orbiter turned away from the nearest HAC until sufficient energy was dissipated to allow a normal approach and landing. The spacecraft slowed to subsonic velocity at approximately 49,000 feet, about 25 miles from the runway.

The approach-and-landing phase began at approximately 10,000 feet at an equivalent airspeed of 320 mph, roughly 8 miles from touchdown. Autoland guidance was initiated at that point to guide the orbiter to the −19-degree glideslope (about seven times as steep as a commercial airliner), aimed at a target 1 mile in front of the runway. The speed brake was used to control velocity, and at 1,750 feet above ground level a preflare maneuver was started to position the spacecraft for a 1.5-degree glideslope in preparation for landing. The final phase reduced the orbiter's sink rate to less than 9 ft/sec. Touchdown occurred approximately 2,500 feet past the runway threshold at a speed of roughly 220 mph.

The National Aero-Space Plane served as a technology driver for the aerospace community throughout the mid-1980s, but the program never got to the flight stage. NASA.

CHAPTER 6

Back to the "Ablative" Future

This chapter explores efforts to develop new reentry and landing concepts from the 1990s on. During this period, a series of ideas emerged that ranged from the DC-X powered landing concept to the return of a metallic heat shield for the National Aero-Space Plane (NASP) and X-33. The chapter includes a discussion of the decision to pursue development of the Crew Exploration Vehicle (CEV) using a capsule, a blunt-body, an ablative heat shield, and parachutes (or perhaps a Rogallo wing) to return to Earth (or perhaps land in the ocean) in 2005, as well as the decision to move toward a commercial piloted spacecraft that could be reusable and could possibly serve as a capsule. The recovery of scientific sample return missions to Earth (both the loss of Genesis and the successful return of Stardust) suggests that these issues are not exclusive to the human space flight community. While the effectiveness of the thermal protection system on the Space Shuttle has been demonstrated, its fragility and high maintenance costs remained an important concern for follow-on space reentry and recovery efforts. As a result, a large slate of new research programs on thermal protection systems for various vehicles—such as NASP, X-33, and X-34—followed in the 1980s and 1990s. Some of those efforts were exotic, but they eventually led full circle to a new version of ablative heat shields developed for the Genesis and Stardust missions, as well as the heat shield's most advanced concept, expressed in the Constellation program.

The Rise (and Fall) of the X-30
National Aero-Space Plane

Many aerospace engineers have long believed that the best solution to the world's launch needs would be a series of completely reusable launch vehicles (RLVs). A debate has raged between those who believe RLVs are the only—or at least the best—answer and those who emphasize the continuing place of expendable launch vehicles (ELVs) in future space-access operations. This debate relates to estimates of the number of missions required: the more missions, the more effective a RLV system becomes. However, estimates of the number of future space flights have consistently been too high, thus favoring the ELV concepts. The same debate took place at the beginning of the Space Shuttle's development. RLV advocates have been convincing in their argument that the only course leading to "efficient transportation to and from the earth" would be RLVs, and they have made the case repeatedly since the late 1960s.[1] Their model for a prosperous future in space is the airline industry, with its thousands of flights per year and its exceptionally safe and reliable operations. Since the advent of the Space Shuttle, NASA has been committed to advancing this model with Shuttle follow-on efforts, until a radical departure in 2005 with the short-lived Constellation program.

One especially important effort involved work on a hybrid air and spaceplane that would enable ordinary people to travel halfway around the globe in about 1 hour. Such a concept was quite simple in theory but enormously complex in reality. It would require developing a passenger spaceplane with the capability to fly on wings from an ordinary runway like a conventional jet. Flying supersonic, it would reach an altitude of about 45,000 feet, at which point the pilot would start scramjet engines—a more efficient, faster jet engine with the potential to reach hypersonic speeds in the Mach 3 realm. These capabilities would take the vehicle to the edge of space for a flight to the opposite side of the globe, where the process would be reversed, and the vehicle would land like a conventional airplane. The vehicle would never reach orbit, but it would still fly in space, and the result would have been the same as orbital flight for passengers but for less time. It would even be possible, RLV supporters insisted, to build such a spaceplane that could reach orbit.[2]

One of the most significant efforts toward developing this reusable spaceplane was the NASP, a joint NASA/Air Force technology demonstrator begun in the early 1980s. With the beginning of the Ronald Reagan administration, and its associated military buildup, Tony DuPont, head of DuPont Aerospace, offered an unsolicited proposal to the Defense Advanced Research Projects Agency (DARPA) to design a hypersonic vehicle powered by a hybrid-integrated engine of scramjets and rockets. DARPA Program Manager Bob Williams liked the idea, and he funded it as a "black" program code-named "Copper Canyon" between 1983 and 1985. The Reagan administration later unveiled it as the

---

1. This was the argument made to obtain approval for the Space Shuttle. See *The Post-Apollo Space Program: A Report for the Space Task Group* (Washington, DC: NASA, September 1969), pp. 1, 6.

2. Fred Hiatt, "Spaceplane Soars on Reagan's Support," *Washington Post,* February 6, 1986, p. A4; Roger Handberg and Joan Johnson-Freese, "If Darkness Falls: The Consequences of a United States No-Go on a Hypersonic Vehicle," *Space Flight* 32 (April 1990): 128–131; Linda R. Cohen, Susan A. Edelman, and Roger G. Noll, "The National Aerospaceplane: An American Technological Long Shot, Japanese Style," *American Economic Review* 81 (1991): 50–53; Roger Handberg and Joan Johnson-Freese, "Pursuing the Hypersonic Option Now More Than Ever," *Space Commerce* 1 (1991): 167–174.

National Aero-Spaceplane, designated the X-30. Reagan called it "a new Orient Express that could, by the end of the next decade, take off from Dulles Airport and accelerate up to twenty-five times the speed of sound, attaining low-Earth orbit or flying to Tokyo within two hours."[3] With this public announcement of NASP, the hypersonic aerospaceplane had returned.[4]

The NASP program initially proposed to build two research craft, at least one of which was intended to achieve orbit by flying in a single stage through the atmosphere at speeds up to Mach 25. The proposed X-30 would use a multicycle engine that shifted from jet to ramjet to scramjet speeds as the vehicle ascended, burning liquid hydrogen fuel with oxygen scooped and frozen from the atmosphere.[5] It never achieved anything approaching flight status.

NASP, like earlier hypersonic flight research projects, fell victim to budget cuts, but this time as a result of the end of the Cold War. With costs approaching $17 billion, some 500 percent over the projected budget, it was also a program 11 years behind schedule when canceled. With cost overruns and delays in scheduling, the project was less attractive than it had been originally. Accordingly, Congress canceled NASP in 1992, during fiscal year 1993 budget deliberations. In 1994, the program died a merciful death, trapped as it was in bureaucratic politics and seemingly endless technological difficulties.[6]

---

3. Ronald Reagan, "State of the Union Address," February 4, 1986.

4. Richard P. Hallion, "Yesterday, Today, and Tomorrow: From Shuttle to the National Aero-Spaceplane," in Richard P. Hallion, ed., *The Hypersonic Revolution: Case Studies in the History of Hypersonic Technology, Volume II, From Scramjet to the National Aero-Space Plane (1964–1986)* (Bolling AFB, DC: USAF Histories and Museums Program, 1998), pp. 1334, 1337, 1340–1341, 1345, 1362–1364; T.A. Heppenheimer, *The National Aerospaceplane* (Arlington, VA: Pasha Market Intelligence, 1987), p. 14; Larry Schweikart, "The Quest for the Orbital Jet: The National Aerospace Plan Program, 1983–1995," manuscript, pp. I.30-I.31, NASA Historical Reference Collection; John V. Becker, "Confronting Scramjet: The NASA Hypersonic Ramjet Experiment," Case VI in Hallion, ed., *The Hypersonic Revolution*, vol. II, pp. vi–xv.

5. Larry Schweikart, "Command Innovation: Lessons from the National Aerospaceplane Program," in Roger D. Launius, ed., *Innovation and the Development of Flight* (College Station, TX: Texas A&M University Press, 1999), pp. 299–322.

6. Roger Handberg and Joan Johnson-Freese, "NASP as an American Orphan: Bureaucratic Politics and the Development of Hypersonic Flight," *Spaceflight* 33 (April 1991): 134–137; Larry E. Schweikart, "Hypersonic Hopes: Planning for NASP," *Air Power History* 41 (Spring 1994): 36–48; Larry E. Schweikart, "Managing a Revolutionary Technology, American Style: The National Aerospaceplane," *Essays in Business and Economic History* 12 (1994): 118–132; Larry E. Schweikart, "Command Innovation: Lessons from the National Aerospaceplane Program," in Roger D. Launius, ed., *Innovation and the Development of Flight* (College Station, TX: Texas A&M University Press, 1999), pp. 299–323; *Defense Daily*, April 17, 1992.

Although the program never came close to building or flying hardware, NASP contributed significantly to the advancement of materials that were either capable of repeatedly withstanding high temperatures (on the vehicle's nose and body) or capable of tolerating repeated exposure to extremely low temperatures (the cryogenic fuel tanks). By 1990, NASP researchers had realized significant progress in titanium aluminides, titanium aluminide metal matrix composites, and coated carbon-carbon composites that were all useful for thermal protection because of their resistance to superheating. Moreover, Government and contractor laboratories had fabricated and tested large titanium aluminide panels under approximate vehicle thermal conditions, and NASP contractors had fabricated and tested titanium aluminide composite pieces.[7] All of this research would have contributed to the realization of NASP's capability to reenter the atmosphere, slow to landing speed, and recover from space flight.

Two areas of research were important in NASP's thermal protection system. First, the vehicle's engineers explored several categories of thermal protection systems, but the most significant area involved convectively (or actively) cooled structures—a method of limiting structural temperatures by circulating a coolant through the hot zones of the vehicle—for the fuselage. This technique had been extensively studied as far back as the 1950s, but NASP advanced it considerably, even though it never yielded a workable system.[8] This active cooling approach applied fluids in a cooling loop to high heat-flux areas of the space vehicle. This, of course, required the vehicle to carry excess weight for the cooling system, and it was never feasible when fluids such as liquid metals (like sodium or potassium) were the coolant. This approach became more tenable when NASP pioneered development of a system that used the liquid hydrogen of the fuel system for this purpose. The weight of the fuel then served a dual purpose in the spaceplane.[9] Ironically, the highest speeds possible in the combined-cycle engines under development for NASP were actually aided through the effective and imaginative use of active thermal management. As late as February 7, 1992, as the program was about to be canceled, a NASP structure was filled with liquid hydrogen (at –423 °F). The assembly then successfully endured bending and heating to 1,300 °F on the shell. This process

---

7. Schweikart, "The Quest for the Orbital Jet," pp. III.37-III.38 and III.41-III.42.

8. H. Neale Kelly and Max L. Blosser, "Active Cooling from the Sixties to NASP," presentation at Current Technology for Thermal Protection Systems Workshop, Hampton, VA, February 11–12, 1992, copy in possession of authors.

9. W. Coleman, T. Dansby, and R. Sheldon, "NASP Technology Option Six, Leading Edge Cooling," NASP-CR-1082 (May 1990).

simulated atmospheric heating as high as Mach 16 and was an important success, but it was not sufficient for orbital flight.[10]

A major challenge, never truly resolved during the NASP program, involved curtailing the mass and weight of the active thermal-management system, which grew ever more complex and massive as speeds increased. Although much of the program was highly classified, this seems to remain a future research question.[11]

Second, in addition to convective cooling systems, the NASP program pursued research in advanced materials, including various composites and titanium-based alloys that retained structural integrity at temperatures sometimes in excess of 1,800 °F.[12] NASP engineers pursued significant advances in metal matrix composites, consisting of several threads of research in advanced metal matrices and high-strength fiber composites. Advanced titanium and beryllium alloys had high strength and high temperature resistance while weight remained low. These lightweight materials possessed good thermal conductivity that would be critical in the successful completion of a single-stage-to-orbit (SSTO) vehicle.[13]

The X-33/VentureStar™ Program

NASA began its own RLV program after the demise of NASP in the mid-1990s, and the Agency's leadership expressed high hopes for the proposed X-33, a small suborbital vehicle that would demonstrate the technologies required for an operational SSTO launcher. This technology demonstrator was intended as the first of a projected set of four stages that would lead to a routine spacefaring capability. The X-33 project, undertaken in partnership

---

10. Charles Morris, "NASP Technology Transfer Presentation to ITP Workshop," NASA Office of Aeronautics and Space Technology (March 18, 1992).

11. "DARPA Chief Notes Potential of Supersonic Combustion Ramjet," *Aerospace Daily*, March 29, 1985; "NASP Moves at Slower Speed," *Military Space*, July 17, 1989, pp. 1, 7–8; United States General Accounting Office, "National Aero-Spaceplane: Key Issues Facing the Program" (March 31, 1992).

12. United States General Accounting Office, "National Aero-Spaceplane: A Technology Development and Demonstration Program to Build the X-30," USGAO/NSIAD-88-122, April 1988, pp. 38–39.

13. Charles Morris, "NASP Technology Transfer Presentation to ITP Workshop," NASA Office of Aeronautics and Space Technology (March 18, 1992); T.M.F. Ronald, "Materials Related to the National Aero-Spaceplane," NASP Joint Program Office Aeronautical Systems Center, Air Force Materiel Command, Wright-Patterson AFB, OH, 1994.

An artist's concept for the X-33. The X-33 program was created in 1997 to develop technologies that were intended to pave the way for a full-scale, commercially developed reusable launch vehicle. NASA 9906365.

with Lockheed Martin, had an ambitious timetable to fly by 2001. But what would happen after its tests were completed remains unclear. Even assuming complete success in meeting its R&D objectives, the time and money necessary to build, test, and certify a full-scale operational follow-on version remains problematic. Who would pay for such an operational vehicle also remains a mystery, especially since the private sector had become less enamored with the joint project over the years and eased itself away from the venture.[14]

There is also an understanding that the technical hurdles have proven more daunting than anticipated, as was the case 30 years ago with the Space Shuttle and more recently with NASP. Any SSTO vehicle (and X-33 held true to this pattern) would require breakthroughs in a number of technologies, particularly in propulsion, materials, and thermal protection. When designers are to begin work on the full-scale SSTO, they might find that available technologies limit payload size so severely that the new vehicle provides little or no cost savings compared to old launchers. If this were to become the case, then everyone must understand that NASA would receive the same type of criticism that it had experienced over the Space Shuttle. Without question, this related to the inability to accurately predict the flight rate for space-access missions.[15]

This is not to say that an SSTO vehicle could never work or that the X-33 should not have been pursued. It has always been NASA's job to take risks and push the technological envelope. However, while the goal may be the development of a launch system that is significantly cheaper, more reliable, and more flexible than presently available, it is possible to envision a future system that cannot meet those objectives. This is all the more true in a situation in which breakthrough technologies had not emerged.[16]

Even so, NASA joined Lockheed Martin in an unusual partnership to develop the X-33, and despite the outcome of that particular effort it might prove a model worth returning to in the 21st century. NASA undertook this effort because over the past two decades the Space Shuttle, a partially reusable vehicle, taught political leaders that bringing down costs and reducing turnaround time were much greater challenges than originally anticipated. Even at the time, it was a conscious decision on the part of NASA and the White House to select a partially reusable Space Shuttle to minimize the development costs, fully recognizing the decision would have an adverse effect on long-term

---

14. Frank Sietzen, "VentureStar Will Need Public Funding," *SpaceDaily Express*, February 16, 1998, NASA Historical Reference Collection.

15. Greg Easterbrook, "The Case Against NASA," *New Republic*, July 8, 1991, pp. 18–24; Alex Roland, "Priorities in Space for the USA," *Space Policy* 3 (May 1987): 104–114; Alex Roland, "The Shuttle's Uncertain Future," *Final Frontier*, April 1988, pp. 24–27.

16. James A. Vedda, "Long-term Visions for U.S. Space Policy," background paper prepared for the Subcommittee on National Security, International Affairs, and Criminal Justice of the House Committee on Government Reform and Oversight (May 1997).

VentureStar® was to have been a follow-on to the smaller, experimental X-33 technology demonstrator, but cost, schedule, and technological problems led to its cancellation in April 2001. NASA MSFC-9906386.

operational costs.[17] The X-33 program was intended to meet those objectives of cost effectiveness and reliability that the Shuttle had left unfulfilled. It was to have served as a technology demonstrator, but upon successful flight Lockheed Martin agreed to develop and fly a commercial version, VentureStar™, in the first decade of the 21st century.

That cooperative agreement, furthermore, was at the heart of NASA's efforts to initiate so-called new ways of doing business by serving as both procurement and management tools. As a new procurement instrument, the cooperative agreement was part of a fast-track managerial approach (and part of the larger "faster-better-cheaper" formula) that featured more expeditious acquisition procedures, streamlined bureaucracy, limited oversight, and lessened

17. On the Space Shuttle as a launch vehicle, see Dennis R. Jenkins, *Space Shuttle: The History of Developing the National Space Transportation System—The First 100 Missions* (Cape Canaveral, FL: 2001); Ray A. Williamson, "Developing the Space Shuttle," in Logsdon, gen. ed., *Exploring the Unknown*, pp. 4:161–191.

investment by NASA. It also mandated that industry share the costs of the project so that both sides shared the risk.[18]

Given the problems experienced on the X-33 program (with delays of more than 1 year because of difficulties with critical elements such as the fuel tanks), some criticized the use of the cooperative agreement as the culprit in sidelining the project. That seems to put too fine a point on the issue, as perhaps the greatest advantage of that cooperative agreement was its fostering of shared responsibility for the funding of launcher development between the public and private sectors. The cooperative agreement also provided industry an important voice in launcher planning.

Before the X-33, the space industry had never expended significant resources in launcher development. The industry contribution to X-33 development was $442 million through 2000. In an era of declining Government space R&D budgets, the importance of that number cannot be underestimated. It seems obvious that although sizable Government investment in the development of future launchers will be required, there is a reason to pursue additional cooperative projects.[19] If a new generation of launchers is to be developed in the first part of the 21st century, it appears that the Federal Government will have to take a leadership role in identifying R&D funds and working in partnership with industry. Perhaps this would be most effectively accomplished through additional cooperative agreements.

Of course, one may also legitimately criticize the overall X-33 effort as a self-deception that a single Government program—even one using a new type of partnership with industry—will be able to solve the problems of space access. In this case, like NASP of the 1980s, an ambitious program was created, hyped as the panacea for all space-access challenges, underfunded, and then ridiculed when it failed. Unrealistic goals, coupled with impossible political demands, left the X-33 program stunned and stunted at century's end. As one space policy analyst concluded, "it continued to blur the line, which

---

18. Stephanie A. Roy, "The Origin of the Small, Faster, Cheaper Approach in NASA's Solar System Exploration Program," *Space Policy* 14 (August 1998): 153–171; Tony Spear, "NASA FBC Task Final Report" (March 2000), pp. 1–2; DC-X Evaluation Team, "DC-X as a Demonstrator for SSTO Technologies," briefing to NASA Administrator, March 1, 1994, pp. 17, 31; and NASA Office of Inspector General, "Review of National Aeronautics and Space Administration Cooperative Agreements With Large Commercial Firms," P&A-97-001 (August 22, 1997), pp. 4–5, all in NASA Historical Reference Collection.

19. Andrew J. Butrica, "The NASA-Industry Cooperative Agreement as a Tool for Reusable Launcher Technology Development," paper presented at the Society for History in the Federal Government annual meeting, Washington, DC, March 16, 2000.

should be bright, between revolutionary, high-risk, high-payoff R&D efforts and low-risk, marginal payoff evolutionary efforts to improve systems."[20] The willingness to embark on this effort has been a mystery to many analysts as they review the linkage— which some called "foolish"—between the high-risk R&D of the X-33 and the supposedly operational VentureStar.

The X-33 program used direct technology developments from the NASP program for its thermal protection system and, in the process, sought to avoid a large-scale R&D program. Using metal and metal matrix composites in thermal protection materials, the X-33 sought to exploit those materials' robustness and durability while sacrificing tolerance for extremely high temperatures. Project engineers refrained from the ceramic tiles of the Space Shuttle in favor of a metallic thermal protection system because of its intrinsic ductility and inherent resistance to water. Metallic thermal protection systems had a long history to this point. The notable metallic heat-sink materials include the copper nose tips on early Jupiter intermediate-range ballistic missiles, the Beryllium nose tips used on the Polaris submarine-launched ballistic missiles, and the Inconel X used on the X-15.[21] Even so, the X-33 engineering team insisted that the new thermal protection system would be pathbreaking: "The success of the X-33 will overcome the ballistic reentry TPS mindset. The X-33 TPS is tailored to an aircraft type mission while maintaining sufficient operational margins. The flight test program for the X-33 will demonstrate that TPS for the RLV is not simply a surface insulation but rather an integrated aeroshell system."[22]

Approaching the X-33 thermal protection system presented engineers with significant challenges. Of course, the lifting body configuration of the X-33 also offered some opportunities. The two principal objectives for the thermal protection system were to shield the primary airframe structure from excessive thermal loads and to provide an aerodynamic surface for the vehicle. Because the hypersonic environment for a lifting body is well understood, the X-33 team chose an integrated standoff aeroshell design with minimal weight to reduce procurement and operational costs. As one research report noted:

---

20. Scott Pace to Roger D. Launius, September 17, 2000, copy in possession of authors.

21. James M. Grimwood and F. Strowd, "History of the Jupiter Missile System," History and Reports Control Branch, U.S. Army Ordinance Missile Command, July 1962; R.D. Clark, J.R. Mullaly, T.A. Wallace, and K.E. Wiedeman, "Emittance/Catalysis/Oxidation Coatings for Titanium-Aluminide Intermetallic Alloys," NASP-TM-1005 (June 1989).

22. S. Bouslog, J. Mammano, and B. Strauss, "Thermal Management Design for the X-33 Lifting Body," p. 1, NASA CR-1998-208247.

Today, metallic TPS found on the X-33 is a hybrid of materials sealed in a panel. For this type of thermal protection system, very thin foils act with a metallic honeycomb and fibrous insulation to form a TPS panel. The approach can provide the durable acreage coverage needed for reusable launch vehicles. The degree of maintenance on these metallic systems should be a fraction of the maintenance that the tiles require on the space shuttle. These metallic systems may require emissive coatings to help cool the vehicle on orbit. If the launch vehicle is orbital, acreage metallic TPS will need a coating with high emissivity in order to cool during orbit. The defunct National AeroSpace Program [sic] (NASP) developed such a coating tailored to titanium aluminide.[23]

The TPS developed for the X-33 consisted of diamond-shaped metallic panels approximately 46 centimeters along each side. Each of these panels incorporated a metallic honeycomb sandwich heat-shield outer panel with foil-encapsulated fibrous insulation attached to the inner side of the heat-shield panel. To put these shingles together on the vehicle, NASA used a metallic standoff rosette at each corner attached, in turn, to a composite support structure.

By 1998, much of the basic design work had been completed on the X-33 thermal protection system, and tests were underway. As one report noted:

> There are two basic panel types, with two material variations on the windward body. The primary panel type is made of Inconel 617 with .006" inner and outer facesheets brazed to .0015" thick, 3/16" cell core with a total thickness at .50". Areas on the vehicle that are subjected to higher temperatures require .010" thick facesheets and .0035", 3/16" cell core. The other primary panel type is a .50" achined isogrid made of Inconel 617, or MA 754 where exposed to higher temperatures. The isogrid panels are used primarily at locations of high curvature where producibility or high stresses dictate their use. Isogrid panels are also used when a penetration, such as an RCS thruster or antenna[,] is located. The use of isogrid panels is minimized, as they are relatively heavier than honeycomb brazed panels....

---

23. Jeffrey D. Guthrie, Brigitte Battat, and Barbara K. Severin, "Thermal Protection Systems for Space Vehicles," *Material Ease*, AMPTIAC 11.

The panel-to-panel seals consist of a primary shingle seal, which is an extension of the outer skin, and a secondary seal system. The secondary seal system has a 'J' hook and a leaf seal that contacts the 'J' seal. The secondary seal system has been designed to provide "fly home" capability if the primary seal fails.[24]

Of course, the thermal protection system was designed to perform the dual role of supporting the aerodynamic-pressure loads and providing thermal protection. The critical question was that, although metallic thermal protection system concepts had been demonstrated previously, could one be developed to protect the spacecraft at orbital reentry speeds and heating.[25]

The metallic thermal protection system of the X-33 originated through years of research sponsored by NASA and the Department of Defense as well as through its manufacture by Rohr Industries. Progressing from early standoff heat shields to multi-wall concepts, thermal protection systems by the latter stages of NASP used state-of-the-art prepackaged superalloy honeycomb sandwich panels. The X-33 advanced this effort and undertook detailed design and fabrication of these superalloy honeycomb thermal protection system elements, finding that they worked relatively effectively during evaluation.[26] Modifications resulting

24. BFGoodrich Aerospace Aerostructures Group, "Annual Performance Report X-33 Thermal Protection System for the Period April 1, 1998–March 31, 1999," pp. 6–7, copy in possession of authors.

25. Randal W. Lycans, "X-33 Base Region Thermal Protection System Design Study," paper presented at 7th AIAA/American Society of Mechanical Engineers Joint Thermophysics and Heat Transfer Conference, June 16–18, 1998; H. Miura, M. Chargin, J. Bowles, T. Tam, D. Chu, and M. Chainyk, "Transient Analysis of Thermal Protection System for X-33 Vehicle Using MSC/NASTRAN," available online at *www.mscsoftware.com/support/library/conf/auc97/p01897.pdf*, accessed October 10, 2009.

26. H.L. Bohon, J.L., Shideler, and D.R. Rummler, "Radiative Metallic Thermal Protection Systems: A Status Report," in *Journal of Spacecraft and Rockets* 12 (October 1977): 626–631; L.R. Jackson and S.C. Dixon, "A Design Assessment of Multiwall, Metallic Stand-off, and RSI Reusable Thermal Protection Systems Including Space Shuttle Application," NASA TM-81780 (April 1980); J.L. Shideler, H.N. Kelly, D.E. Avery, M.L. Blosser, and H.M. Adelman, "Multiwall TPS—An Emerging Concept," *Journal of Spacecraft and Rockets* 19 (July–August 1982): 358–365; W. Blair, J.E. Meaney, and H.A. Rosenthal, "Re-Design and Fabrication of Titanium Multi-Wall Thermal Protection System (TPS) Test Panels," NASA CR-172247 (January 1984); W. Blair, J.E. Meaney, and H.A. Rosenthal, "Fabrication of Prepackaged Superalloy Honeycomb Thermal Protection System (TPS) Panels," NASA CR-3755 (October 1985); J. Anderson, R.B. LeHolm, J.E. Meaney, and H.A. Rosenthal, "Development of Reusable Metallic Thermal Protection System Panels for Entry Vehicles," NASA CR-181783 (August 1989); M.P. Gorton, J.L. Shideler, and G.L. Web, "Static and Aerothermal Tests of a Superalloy Honeycomb Prepackaged Thermal Protection System," NASA TP-3257 (March 1993).

from tests of the system led to the outer honeycomb surface supported at four points by an underlying latticework that formed the aeroshell, supporting the vehicle's aerodynamic shape.[27] As one study of this effort noted:

> Fibrous insulation, beneath the outer honeycomb sandwich panel, is encapsulated by foil attached to the perimeter of the panel. Seals between panels on the outer surface are intended to maintain the aerodynamic pressure across the outer, hot honeycomb sandwich. In contrast, the prepackaged superalloy honeycomb TPS is vented so that there is little pressure difference across the outer honeycomb sandwich when it is hot.[28]

Despite relative success with the X-33 thermal protection system, the X-33 was mired in seemingly inscrutable technological problems and bureaucratic challenges, and NASA lost faith in it and terminated the effort in 2001. Thereafter, NASA officials expressed a deeper understanding that the technical hurdles proved more daunting than anticipated, as was the case 30 years ago with the Space Shuttle, and more recently with the NASP.

Any successful SSTO vehicle (and the X-33 program reinforced this pattern) would require breakthroughs in a number of technologies, particularly in propulsion and materials.[29] Some engineers referred to X-33 as being built from "unobtainium," and they thought the United States should instead pursue more conventional space-access technologies. Without going that far, many engineers recognize four major challenges that complicate efforts to develop this technology:

1. Aerodynamics
2. Guidance and Control
3. Materials
4. Propulsion

In the first realm of aerodynamics, through a succession of projects and studies, researchers have overcome many of the roadblocks to create effective shapes for a hypersonic vehicle. Many of the aerodynamic questions are now

---

27. Leonard David, "X-33 Flight Test Vehicle to Use Metallic Panels," *Space News,* September 23–29, 1996.

28. Max L. Blosser, "Development of Metallic Thermal Protection Systems for the Reusable Launch Vehicle," NASA TM-110296 (October 1996), pp. 2–3.

29. Leonard David, "NASA Shuts Down X-33, X-34 Programs," *Space.com,* March 1, 2001, available online at *http://www.space.com/missionlaunches/missions/x33_cancel_010301.html,* accessed March 28, 2003.

satisfactorily understood. The same is true for the second challenge of guidance and control. Materials research remains an important aspect yet to be resolved as research continues on heat-resistant materials and composites that can reduce weight. But the biggest issue remains propulsion. There is, as yet, no fully functional scramjet engine (at least not in the public arena). Furthermore, scramjet engines require high dynamic pressure to operate effectively and are not ideally suited to the SSTO mission, which requires acceleration at low and zero dynamic pressure. These are all problems for future researchers to solve before realizing the dream of a true SSTO vehicle.[30]

X-34 and the Continuing Challenge of Reentry

Almost in parallel with the X-33 program, NASA undertook the X-34—also known as the Reusable Small Booster Program—to demonstrate certain technologies and operations useful to smaller reusable vehicles launched from an aircraft. Among those technologies were autonomous ascent, reentry, and landing; composite structures; reusable liquid-oxygen tanks; rapid vehicle turnaround; and thermal protection materials. The X-34 was considerably smaller and lighter than the X-33 and was supposed to be capable of hypersonic flight to Mach 8—but not nearly as capable as the X-33's Mach 15. Consequently, the X-34 was to have been much less expensive and simpler to develop, operate, and modify.[31]

Design of the thermal protection system for this hypersonic flight vehicle required (just as it had for the X-33) determining the peak temperatures over the surface and the heating-rate history along the flight profile. Based on these determinations, project engineers employed a system that would work for the suborbital technology demonstrator containing advanced thermal protection systems capable of surviving subsonic flights through rain and fog. As one report stated:

> A passive system, such as that employed on the X-34, takes advantage of the insulative properties of the TPS materials to hold a significant portion of the incident heat until it is radiated or convected away from the vehicle. The two critical parameters that

---

30. Roger D. Launius, "Hypersonic Flight: Evolution from X-15 to Space Shuttle," AIAA-2003-2716, delivered at "Next Century of Flight" conference, Dayton, OH, July 2003.

31. John W. Cole, "X-34 Program," in "X-33/X-34 Industry briefing, 19 October 1994," especially p. 1A-1,216, NASA Historical Reference Collection.

Artist concept of X-34 technology demonstrator, another second-generation RLV, in flight. The X-34 was a reusable-technology test that was to be capable of speeds up to Mach 8 and altitudes of 250,000 feet. NASA 9906366.

determine the TPS options suitable for a given vehicle and trajectory are the peak-heating rate and the integrated heating over the time of the flight profile. The former determines the maximum temperature environment, and thus the material options, and the latter determines the thickness distribution of the insulative TPS material required to protect the surface. Except for the nose-cap and wing leading edges [where silicone impregnated reusable ceramic ablator (SIRCA) tile is used], the X-34 is protected with flexible-blanket or felt systems.[32]

There was very little subsequent application of this research effort because the project was canceled in 2001, and it did not progress to the flight stage.[33]

---

32. Kathryn E. Wurster, Christopher J. Riley, and E. Vincent Zoby, "Engineering Aerothermal Analysis for X-34 Thermal Protection System Design," *Journal of Spacecraft and Rockets* 36 (March–April 1999): 216–228.

33. F.S. Milos and T.S. Squire, "Thermostructural Analysis of X-34 Wing Leading-Edge Tile Thermal Protection System," *Journal of Spacecraft and Rockets* 36 (March–April 1999): 189–198.

X-38 Recovery Parafoil

In a departure from the X-33/X-34 thermal protection system story, NASA also pursued parafoil development in the late 1990s as a dedicated method of returning the crew aboard the ISS to Earth. Although through 2009 the ISS could handle only a crew of three, when it was completed an international crew of up to seven was intended to be able to live and work in space for between 3 and 6 months. Space agencies anticipated that a major problem in sending up a large crew, however, would be returning them safely to Earth in the event of an emergency. This led to a contentious debate over the value of the Space Station, as funding and program schedule again came under fire from critics. To ensure the crew size of seven, in 1994 NASA added to the program a U.S.-developed Crew Return Vehicle (CRV). Even so, it requested no additional funding to pay for the CRV. NASA carried the CRV as an overage in the budget for ISS until funding was finally allocated in the fiscal year 1999 budget submission to Congress. But that was only a small part of the problem.

As always, NASA engineers began the X-38 CRV project with optimism. The Agency had a record of excellence that had been forged during Project Apollo in the 1960s, and success after success had followed, including landings on Mars and voyages to the outer planets of the solar system. A culture of competence permeated NASA. Actor Robert Guillaume voiced the beliefs of many when in a sitcom he said, "You put an X anyplace in the solar system, and the engineers at NASA can land a spacecraft on it."[34] Nothing more effectively stated the public conception of NASA's culture of competence than this public announcement. Unfortunately, that optimism and skill was misplaced in the case of the X-38, as costs and schedule combined to defeat the program.

The X-38 program really began at Dryden Flight Research Center in 1992, when Dale Reed, a veteran of the lifting body research of the 1960s and 1970s, began work for the JSC to test a CRV concept modeled on those earlier efforts. This proved successful, and in-house development of the X-38 concept began at JSC in early 1995. In the summer of 1995, early flight tests were conducted of the parafoil concept, dropping platforms with it from an aircraft at the Army's Yuma Proving Ground, Yuma, AZ. In early 1996, NASA awarded a contract to Scaled Composites, Inc., of Mojave, CA, for the construction of three full-scale atmospheric test airframes. In September 1996, the first vehicle airframe was delivered to JSC, where it was outfitted with avionics, computer systems, and other hardware in preparation for flight tests at Dryden. The second vehicle was delivered to the Johnson Space Center in December 1996.

---

34. "The Sweet Smell of Air," *Sports Night*, first aired January 25, 2000.

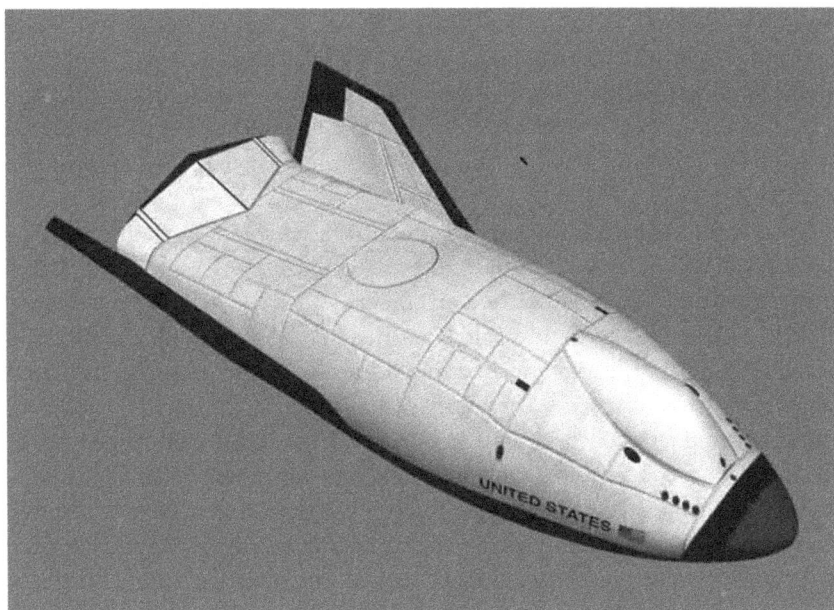

Artist's concept of the X-38 Crew Return Vehicle. Intended to take the place of the Russian Soyuz capsule, the X-38 Crew Return Vehicle was pursued in the late 1990s as an escape vehicle for astronauts and cosmonauts aboard the International Space Station. NASA 9906387.

Testing progressed to an unpiloted space flight test in late 2000. About one hundred people quickly went to work on the project at Dryden and JSC.[35]

The wingless X-38 CRV, when operational, had been intended as the first reusable human spacecraft built in more than two decades. It was designed to fit the unique needs of a Space Station "lifeboat"—long term, maintenance-free reliability always in "turnkey" condition—ready to provide the crew a quick, safe trip home under any circumstances.[36] The starting configuration for the X-38 was the SV-5, an Air Force effort developed under the START

---

35. R. Dale Reed with Darlene Lister, *Wingless Flight: The Lifting Body Story* (Washington, DC: NASA SP-4220, 1997), pp. 187–191; NASA Dryden Flight Center, "X-38 CRV," February 13, 2001, available online at *http://www.dfrc.nasa.gov/Projects/X38/intro.html*, accessed March 31, 2002; NASA Fact Sheet, "The X-38: Low-Cost, High-Tech Space Rescue," IS-2000-01-ISS022-JSC (2000), NASA Historical Reference Collection; Eckart D. Graf, "ESA and the ISS Crew Return Vehicle," *On Station*, March 2001, pp. 3–5.

36. On the lifting body program, see Reed with Lister, *Wingless Flight*; Milton O. Thompson and Curtis Peebles, *Flying Without Wings: NASA Lifting Bodies and the Birth of the Space Shuttle* (Washington, DC: Smithsonian Institution Press, 1999).

program. The SV-5 is the only lifting body shape that has been successfully demonstrated from orbital conditions (X-23 PRIME) to transonic and horizontal landing (X-24A). The NASA Flight Research Center supported the X-24A flight testing, which was treated much as an extension of the NASA lifting body program. During its development effort, the X-38 evolved the basic SV-5 shape.

Data from the aerodynamic studies contributed to the design and operational profile of the Space Shuttle and reemerged in the CRV program. When operational, the CRV was to serve as an emergency vehicle to return up to seven ISS crewmembers to Earth. It would be carried to the Space Station in the cargo bay of a Space Shuttle and attached to a docking port. If an emergency arose that forced the ISS crew to leave the Space Station, the CRV could be boarded, undocked, and after a deorbit engine burn, return to Earth much like a Space Shuttle. Not a true spacecraft in the traditional sense, the vehicle's life support system could sustain a crew for about 7 hours. It used enhanced ceramic matrix components based on carbon-fiber-reinforced carbon, applied as nose and panel thermal protection systems and as hot-structure control surfaces (for the vehicle body), including the roller bearing assembly. Once the vehicle reached the denser atmosphere at about 40,000 feet, a steerable parafoil would be deployed to carry it through the final descent and landing. It would also be fully autonomous, in case the crew was incapacitated or otherwise unable to fly it, so it could return to the landing site using onboard navigation and flight control systems. Backup systems would, of course, allow the crew to pick a landing site and steer the parafoil to a landing.[37]

The X-38 design closely resembled the X-24A lifting body flown at Dryden between 1969 and 1971. That vehicle's body shape, like the CRV, generated aerodynamic lift, which was essential to flight in the atmosphere. The 28 research missions flown by the X-24A helped demonstrate that hypersonic vehicles returning from orbital flight could be landed on conventional runways without power.

---

37. NASA Dryden Flight Research Center Fact Sheet, "X-38," FS-2000-04-038 DFRC (April 2000), NASA Historical Reference Collection; Barton C. Hacker, "The Gemini Paraglider: A Failure of Scheduled Innovation, 1961–64," *Social Studies of Science* 22 (spring 1992): 387–406. The paraglider was conceived during the 1950s as a lightweight hybrid of a parachute and an inflated wing that might allow astronauts to pilot spacecraft to airfield landings. From 1961 to 1964, NASA sought to convert the idea into a practical landing system for the Gemini spacecraft. The spacecraft would carry the paraglider safely tucked away through most of a mission. Only after reentering the atmosphere from orbit would the crew deploy the wing. Having converted the spacecraft into a makeshift glider, the crew could fly to an airfield landing. The system was later further developed, and by the 1990s it was compatible with spaceflight.

An X-38 test vehicle glides down under a giant parafoil toward a landing on Rogers Dry Lake near NASA's Dryden Flight Research Center during its first free flight on November 2, 2000. The huge 7,500-square-foot parafoil enabled the vehicle to land in an area the length of a football field. NASA EC00-0317-41.

Interestingly, the horizontal landing capability demonstrated by the X-24A program was not incorporated into the X-38 program. Instead, it used a steerable parafoil concept for the actual landing. The three prototype X-38s used in the atmospheric flight-testing program were 24.5 feet long, 11.6 feet wide, and 8.4 feet high—approximately 80 percent of the planned size of the CRV. The prototypes were designated V131, V132, and V131R. The V131 prototype was modified for additional testing beginning in the summer of 2000 and thereafter carried the designation V131R. A fourth prototype that was never completed, V133, was intended to incorporate the exact shape and size of the planned CRV. The prototype vehicles were manufactured by Burt Rutan and Scaled Composites in Mojave, CA. These vehicles had shells made of composite materials, such as fiberglass and graphite epoxy, and were strengthened with steel and aluminum at stress points; they weighed between 15,000 and 25,000 pounds.[38]

Another model that was never completed was a fully space-rated X-38 CRV prototype, numbered V201, designed to test the concept under operational conditions. Its inner compartment, representing the crew area, would be a pressurized aluminum chamber. A composite fuselage structure would enclose the chamber, and the exterior surfaces would be covered with a thermal protection system designed to withstand the heat generated by air friction on atmospheric reentry. The thermal protection system would be similar to materials used on the Space Shuttle, but much more durable; it would use carbon and metallic-silica tiles for the hottest regions and flexible blanketlike material for areas receiving less heat during atmospheric reentry.[39]

Much of the technology for the X-38 was to be off the shelf to avoid lengthy R&D efforts. For instance, the flight control computer and the flight software operating system were commercially developed and used in many aerospace applications. Inertial navigation and global positioning systems, similar to units used on aircraft throughout the world, would be linked to the vehicle's flight control system to steer the vehicles along the correct reentry path. Using global positioning already programmed into the navigation system, the flight control computer would become the autopilot that flies the vehicle to a predetermined landing site. Other avionics, flight control systems, materials, and aerodynamics used well-proven concepts and equipment. Finally, the U.S. Army originally

---

38. Ibid.; Reed with Lester, *Wingless Flight*, pp. 131–143, 167–170; NASA Dryden Flight Research Center Press Release, "X-38 Crew Return Vehicle Prototype Resumes Flight Tests," June 28, 2001, NASA Historical Reference Collection.

39. NASA Dryden Flight Research Center Fact Sheet, "X-38," FS-2000-04-038 DFRC (April 2000); NASA Fact Sheet, "The X-38: Low-Cost, High-Tech Space Rescue"; Graf, "ESA and the ISS Crew Return Vehicle," p. 5.

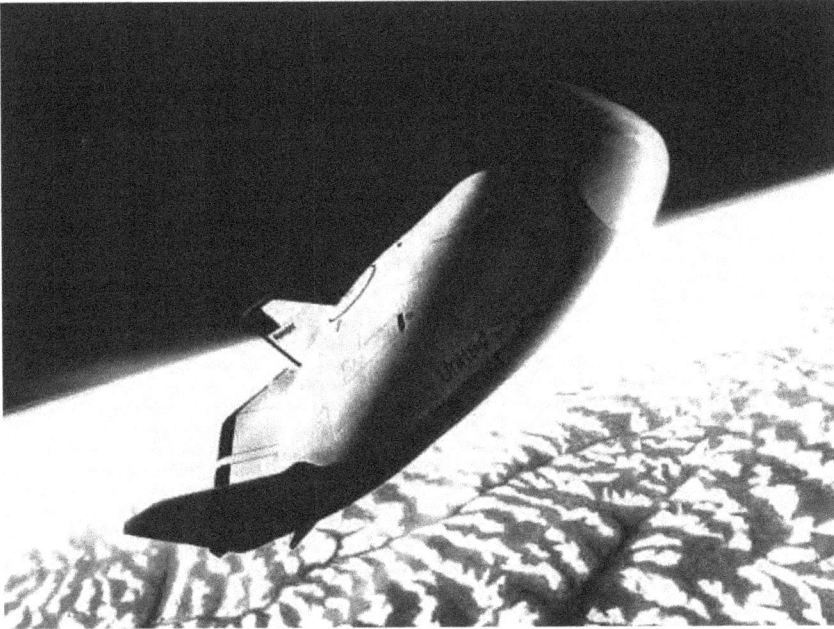

An artist's depiction of NASA's proposed X-38 Crew Return Vehicle reentering Earth's atmosphere. The essential X-38 shape was derived from the earlier X-24A lift body tested at the Flight Research Center during the 1960s. The X-38 test vehicles were at 80 percent of full-size (28.5 feet long and 14.5 feet wide) and weighed approximately 16,000 pounds, on average. In March 2000, Vehicle 132 completed its third and final free flight in the highest, fastest, and longest X-38 flight to date. It was released at an altitude of 39,000 feet and flew freely for 45 seconds, reaching a speed of over 500 mph before deploying its parachutes for a landing on Rogers Dry Lakebed. NASA ED97-43903-1.

developed the design of the parafoil that would deploy in the atmosphere and carry the X-38 to Earth. The European Space Agency (ESA) agreed to develop some of the X-38 systems, developing 15 subsystems or elements of the V201 X-38 spacecraft, scheduled for launch by Space Shuttle Columbia in September 2002. In addition, ESA provided the Guidance, Navigation, and Control (GNC) software for the parafoil phase of the V131R and V133 aerodynamic drop-test vehicles as well as for the supporting tests using a parafoil microlight aircraft and large drop pallets flying the full parafoil.[40]

---

40. NASA Dryden Flight Research Center Fact Sheet, "X-38," FS-2000-04-038 DFRC, April 2000; NASA Fact Sheet, "The X-38: Low-Cost, High-Tech Space Rescue"; Hacker, "The Gemini Paraglider," pp. 387–406; NASA JSC Press Release, "NASA X-38 Team Flies Largest Parafoil Parachute in History," February 3, 2000, NASA Historical Reference Collection; Graf, "ESA and the ISS Crew Return Vehicle," p. 5.

The X-38's thermal protection system development effort benefited significantly from ESA's research and construction of the ceramic nose cap. As one study noted, "Over 10-years of experience in manufacturing carbon-based fiber-ceramic as well as the long-lasting competence in the range of system design of hot structures and thermal protection systems were converted into an innovative component concept. If the past work in this area had a rather experimental character (e.g. co-flight of ballistic capsule missions FOTON and CETEX on EXPRESS), the nose cap as a primary structure component had a mission-crucial task."[41] The study continued with the following:

> During reentry into the earth's atmosphere the nose cap of X-38 experiences the highest thermal load of the entire thermal protection system due to its exposed location within the stagnation point of the vehicle. During the approx. 20 minutes reentry phase surface temperatures of up to 1,750°C were expected with a stagnation pressure of up to 10 to 150 hPa.
>
> The nose cap represents an absolute novelty regarding technological requirements. A component out of fiber-ceramic for thermal loads of up to approx. 1750°C had never been planned before with the claim of re-usability. The nose structure of the American Space shuttle is made of carbon/carbon (C/C) for example and the temperatures reach only scarcely 1500°C.
>
> The connection of the nose shell is made by 8 single fittings, which are made of fiber-ceramic as well respectively in the cooler range of a high temperature-steady (until approx. 1200°C) metal alloy (PM 1000 of Plansee, Austria). The special arrangement and design of the lever-like attachment system guarantees on the one hand a very good mechanical maximum load and on the other hand allows for an unrestricted thermal expansion of the structure, which can amount to 3mm with a medium diameter of the shell of 700 mm and the expected temperature level. In case of a rigid attachment the shell would be destroyed only by the thermal compressive forces.
>
> For thermal isolation a multilevel flexible felt isolation made of oxide fibre-ceramic such as alumina, is intended between the nose shell and the sub-structure. The isolation reduces the maximum temperature at the aluminum sub-structure to about 100 degrees Celsius over a thickness of approx. 45 mm. This isolation

---

41. "Nose Cap of the CRV Test Plant X-38," available online at *http://www.dlr.de/bk/en/desktopdefault. aspx/tabid-4520/7396_read-10111/*, accessed October 12, 2009.

system was developed and manufactured by ASTRIUM. The
entire nose system has a mass of approx. 13.2 kg, whereby the
nose shell alone accounts for 7 kg.[42]

The cap could withstand temperatures of up to 1,750 °C, and by the end
of January 2001, the X-38's nose cap had been delivered to JSC, where it lan-
guished when the program was canceled.

Although the CRV used proven technology for most of its systems, it quickly
became mired in cost and schedule problems. The technical complexity of the
development task, the international character of the effort, and the time phase
for completion all conspired to push costs beyond acceptable bounds. Original
estimates to build a capsule-type CRV amounted to more than $2 billion in
total development cost. After paring down its efforts, the X-38 program team
arrived at a cost estimate of $1.3 billion, but only with several development
and acquisition assumptions yet to be verified. A review of the ISS program
costs in the fall of 2001 noted that for the CRV and other elements, "There is
inadequate current costing information associated with the non-U.S. components." It also found that "Project interruptions will have cost impact on all of
the elements under consideration."[43]

By the spring of 2001 the CRV had been all but eliminated, not so much
because of its cost overruns (although that did contribute), but because of
huge overruns throughout the ISS program. Looking for places to trim the ISS
program budget, the CRV proved an easy target. NASA Administrator Daniel
S. Goldin testified to Congress in April 2001:

> However, the U.S. CRV has a significant set of design activities to
> accomplish before we are ready to enter into a production contract.
> Just last year, NASA's Integrated Action Team, focusing on program
> management excellence, concluded that technology risk reduction
> programs and design definition must be concluded before commit-
> ting to production contracts to best insure that cost, schedule and
> technical targets can be realized. Given the magnitude of planned
> funding dedicated to CRV and the remaining definition work,
> funding allocated for the CRV production phase has been redi-
> rected to help resolve ISS core content budgetary shortfalls. NASA
> has initiated discussions with the European Space Agency (ESA) on

---

42   Ibid.

43. "Report by the International Space Station (ISS) Management and Cost Evaluation (IMCE) Task Force to
     the NASA Advisory Council" (November 1, 2001), p. 7, NASA Historical Reference Collection.

a role in the CRV project. Critical efforts such as X-38 atmospheric flight testing and some preliminary CRV design work and linkages with CTV under SLI will continue so as to maintain viable options for future CRV development. The planned space flight test of the X-38 is under review as part of the program assessment.[44]

When asked at the time if NASA was giving up the CRV, Goldin was emphatic that the CRV was necessary to the Station's completion. As reported by Keith Cowing, "[Goldin] said that NASA was pursuing various options with the international partners on the program for these capabilities. As far as the CRV's precursor, the X-38, Goldin said that there were funds to take that system up to and through orbital testing and that this would be done before NASA makes any commitment on funding the development of a CRV system."[45] But the newly arrived George W. Bush administration directed NASA to halt further development of the CRV. The CRV could have provided the ISS with a permanent replacement for the three-person Soyuz spacecraft currently being used as an interim CRV.[46] NASA still sought to find ways to secure a CRV without direct Agency funding, particularly by having one or more of the international partners take it on, but the results yielded nothing. To this day, the ISS still relies on Russian-built Soyuz capsules, each of which could return a crew of three to Earth.

Bringing Deep Space Probes Back to Earth: Genesis and Stardust

With the diversion of the X-33, X-34, and X-38 lifting reentry efforts, NASA also pursued ablative technology for two small reentry vehicles for science sample collection and return to Earth. The first was the Genesis mission, an effort to collect particles from the solar wind and return them to Earth for laboratory study. Launched on August 8, 2002, atop a Delta II rocket, this spacecraft traveled to a point about 1 million miles from Earth, where it entered a halo orbit around the L1 Lagrangian point between Earth and the Sun. In

---

44. Statement of Daniel S. Goldin, Administrator, NASA, before the House of Representatives Committee on Science, April 4, 2001, NASA Historical Reference Collection.

45. Keith Cowing, "NASA's FY 2002 Budget: Challenges and Opportunities," April 9, 2001, *SpaceRef.com*, online at *http://www.spaceref.com/news/viewnews.html?id=318*, accessed March 31, 2002.

46. Frank Seitzen, Jr., and Keith Cowing, "Habitation Use May Rescue Struggling Commercial Module Project," *SpaceRef.com*, March 27, 2001, NASA Historical Reference Collection.

SRC Recovery Profile

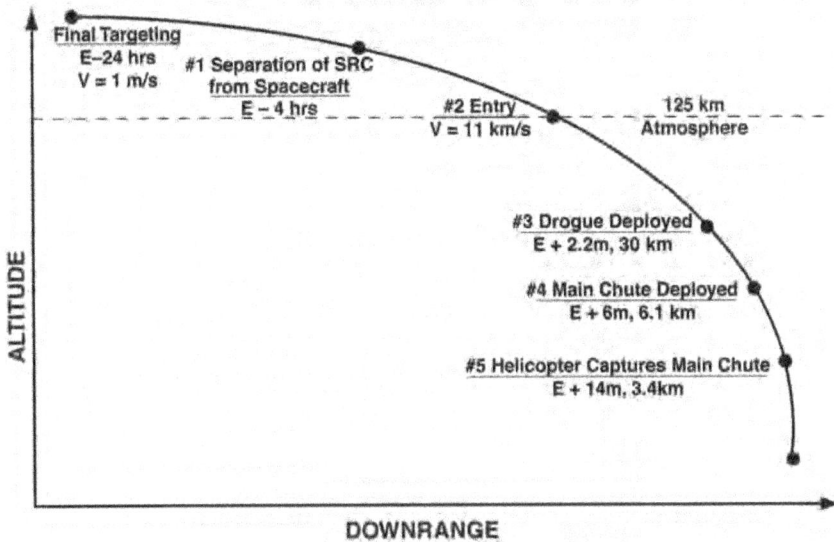

This graph shows the events that were to occur during the recovery of the Genesis sample return capsule in September 2004. The sample return capsule separated from the remainder of the spacecraft about 4 hours before it encountered Earth's atmosphere. Friction initially slowed the capsule, followed by deployment of two parachutes, a small drogue chute, and then a larger main parachute. The parachutes malfunctioned, and two helicopters waiting to grab the capsule's parachute in midair were unable to complete the recovery. The midair catch was to prevent the capsule from coming into contact with Earth materials that could contaminate its cargo of solar wind samples. Unfortunately, it did not work that way. NASA.

April 2004, after 2 ½ years of collecting solar particles, Genesis left the L1 halo orbit. After a flyby of the Moon, Genesis released its return capsule and returned to the L1 halo orbit. The Genesis return capsule entered Earth's atmosphere on September 8, 2004, but the parachute failed to deploy, causing the capsule to impact the ground at a high velocity at the Utah Test and Training Range. Even so, the majority of the science goals were later met.[47]

The ablative heat shields of early space missions—especially Mercury, Gemini, and Apollo—were built specifically for the proposed use, were tailored to a wide

---

47. M. Lo, B. Williams, W. Bollman, D. Han, Y. Hahn, J. Bell, E. Hirst, R. Corwin, P. Hong, K. Howell, B. Barden, and R. Wilson, "Genesis Mission Design," AIAA 98-4468 (August 10–12, 1998); D.S. Burnett, B.L. Barraclough, R. Bennett, M. Neugebauer, L.P. Oldham, C.N. Sasaki, D. Sevilla, N. Smith, E. Stansbery, D. Sweetnam, and R.C. Wiens, "The Genesis Discovery Mission: Return of Solar Matter to Earth," *Space Science Reviews* 105 (January 2003): 509–534.

array of missions scenarios, and were reengineered for each new flight regime. But since the Viking Lander program to Mars in the mid-1970s, NASA had ceased efforts to develop new ablative systems in favor of reusable TPS supporting the Space Shuttle. Accordingly, as one analysis commented:

> As an example, the Pioneer Venus and Galileo missions employed fully dense carbon phenolic that was developed by the United States Air Force for ballistic missile applications. Over the past 30 years NASA adopted a "risk averse" philosophy relative to TPS, i.e., use what was used before, even if it was not optimal, since it had been flight-qualified. An unintended consequence was that the ablative TPS community in the United States slowly disappeared.
>
> NASA completed a number of relatively simple missions using TPS materials developed largely during the 1950s and 1960s. This began to change with the Stardust mission since it used a new ablative TPS because proven technology was not sufficient to ensure the safety of the spacecraft during reentry from deep space.[48]

As intended, the Genesis return to Earth mission was an extension of earlier efforts for a sample return from superorbital missions. One study laid out the manner in which the mission was to unfold as follows:

> About four hours before Earth entry, the spacecraft reorients to the sample return capsule release attitude, spins up to 15 rpm and releases the capsule. Soon after release, the spacecraft reorients to joint its thrusters to Earth and performs a maneuver which will cause the spacecraft to enter Earth's atmosphere, but break-up over the Pacific Ocean. Following release from the spacecraft, the Genesis sample return capsule experiences a passive, spin-stabilized aero-ballistic entry, similar to that of the Stardust mission. When the capsule has decelerated to 1.4 times the speed of sound, the on-board avionics system fires a mortar to deploy the drogue parachute. The drogue is a disk-gap-band design, with heritage dating to the Viking program, and an extensive history of supersonic applications. It serves both to increase the deceleration of the capsule, and to stabilize it through the transonic phase. As the capsule descends into the airspace of the Utah Test & Training Range (UTTR),

---

48. Bernard Laub, "Development of New Ablative Thermal Protection Systems (TPS)," available online at *http://asm.arc.nasa.gov/full_text.html?type=materials&id=1*, accessed October 14, 2009.

The Genesis return capsule entered Earth's atmosphere on September 8, 2004, but the parachute failed to deploy, causing the capsule to impact the ground at a high velocity at the Utah Test and Training Range. This is the Genesis return capsule after impact. Despite the apparent damage, the majority of the science goals were later met. NASA.

> recovery helicopters are directed to fly toward the intercept point. The capsule's ballistic path is designed for delivery within an 84 × 30 km footprint with subsequent reduction to a 42 × 10 km helicopter zone.... The first helicopter on-site will line up and match descent rate, then execute a Mid-Air Retrieval (MAR) capture. If a pass is aborted for any reason the pilot can line up and repeat.[49]

If only it had worked as intended. Instead, Genesis crash-landed in the Utah desert.

While this study is about return to Earth, the story of the Genesis and Stardust thermal protection systems was informed by a major effort to develop the entry vehicle that Galileo carried into Jupiter. Undertaken in the late 1980s and early 1990s, development of the Galileo probe's thermal protection system proved taxing. Bernard Laub concluded the following:

> The Galileo probe to Jupiter was the most challenging entry mission ever undertaken by NASA. The probe employed a 45 deg

---

49. Lo et al., "Genesis Mission Design," pp. 3–4.

blunt cone aeroshell and it entered the Jovian atmosphere at a velocity of ≈ 47.4 km/s. The forebody TPS employed fully dense carbon phenolic (= 1450 kg/m^3) that, at the time, was the best ablator available. The entry environment was very severe and estimates of the peak heating (combined convective and radiative) were on the order of 35 kW/cm2 with a total integrated heat load of ≈ 200 kJ/cm^2.

Because of this experience, NASA engineers concluded that using new materials to replace the carbon phenolic ablative thermal protection systems was appropriate for future missions.[50]

The heat shield for the Genesis return capsule used a carbon-bonded carbon fiber (CBCF) insulating material as a base, an approach used with success in several other NASA missions. The Genesis thermal protection system also included phenolic-impregnated carbonaceous ablator (PICA) and toughened unipiece fibrous-reinforced oxidation-resistant composite (TUFROC). As one study noted:

> CBCF utilized in the above mentioned TPS systems is an attractive substrate material because of its low density and high porosity, superior thermal performance, and compatibility with other components. In addition, it is low cost because of the commercial market it also serves. However, the current CBCF manufacturing process does not produce materials engineered to the specifications NASA desires to put in place. These emerging and highly innovative TPS designs require material manufactured to specification.... The benefits derived include significantly improved flexibility for the TPS design engineer, as well as, more cost efficient CBCF derived TPS fabrication.[51]

As finally flown, the Genesis return capsule was an amalgam of existing technologies and newer concepts. Its backshell was made of SLA-561V "Mars" ablator in an aluminum-honeycomb substructure. Its main structure possessed a carbon-carbon skin with a continuous surface (multilayer fabric layup).[52] As reported by NASA:

50. Ibid.

51. Fiber Materials, Inc., "Advanced Thermal Protection Systems (ATPS), Aerospace Grade Carbon Bonded Carbon Fiber Material," phase-i contract number: NNA05AC11C, 2004.

52. Bill Willcockson, "Genesis Sample Return Capsule Overview," August 25, 2004, Lockheed Martin Space Systems, Denver, CO.

The heat shield is made of a graphite-epoxy composite covered with a thermal protection system. The outermost thermal protection layer is made of carbon-carbon. The capsule heat shield remains attached to the capsule throughout descent and serves as a protective cover for the sample canister at touchdown. The heat shield is designed to remove more than 99 percent of the initial kinetic energy of the sample return capsule.

The backshell structure is also made of a graphite-epoxy composite covered with a thermal protection system: a cork-based material called SLA-561V that was developed by Lockheed Martin for use on the Viking missions to Mars, and have been used on several missions including Genesis, Pathfinder, Stardust and the Mars Exploration Rover missions. The backshell provides the attachment points for the parachute system, and protects the capsule from the effects of recirculation flow of heat around the capsule....

The parachute system consists of a mortar-deployed 2.1-meter (6.8-foot) drogue chute to provide stability at supersonic speeds, and a main chute 10.5 by 3.1 meters (about 34.6 by 12.1 feet).

Inside the canister, a gas cartridge will pressurize a mortar tube and expel the drogue chute. The drogue chute will be deployed at an altitude of approximately 33 kilometers (108,000 feet) to provide stability to the capsule until the main chute is released. A gravity-switch sensor and timer will initiate release of the drogue chute. Based on information from timer and backup pressure transducers, a small pyrotechnic device will cut the drogue chute from the capsule at about 6.7 kilometers altitude (22,000 feet). As the drogue chute moves away, it will extract the main chute. At the time of capture, the capsule will be traveling forward at approximately 12 meters per second (30 miles per hour) and descending at approximately 4 meters per second (9 miles per hour).[53]

On September 8, 2004, during reentry, the thermal protection system worked well, but the parachute system failed and the Genesis return capsule streaked into the ground at the Utah Test and Training Range, southwest of Salt Lake City. A mortar should have fired at 100,000 feet, releasing a drogue parachute to slow and stabilize the vehicle. Then, at 20,000 feet the drogue would have pulled out the main parafoil, and the spacecraft would slow to 9 mph, whereupon

---

53. "Genesis: Search for Origins," June 24, 2008, available online at *http://genesismission.jpl.nasa.gov/gm2/spacecraft/subsystems.html*, accessed October 20, 2009.

helicopters could capture the capsule from midair using a hook. However, the helicopter missed it and the capsule plowed into the ground. The 5-foot-wide, 420-pound capsule burrowed into the desert floor at a speed of 193 mph. In the end, NASA scientists were able to recover some of the solar-wind particles from the Genesis sample return capsule, and they continue to analyze results from this data.[54]

The Stardust sample return mission turned out somewhat better, at least in the effectiveness of the reentry and recovery system. Stardust was launched in 1999 and was recovered in 2006, and it was the first U.S. space mission dedicated solely to returning extraterrestrial material from beyond the Moon. It collected samples from Comet Wild 2 and from interstellar dust. The Stardust return system had six major components: a heat shield, back shell, sample canister, sample collector grids, parachute system, and avionics. The canister was sealed in an exterior shell that protected the samples from the heat of reentry. Scientists believe the material Stardust returned could date from the formation of the solar system, and scientific studies of the samples may alter our understanding of the universe. One major discovery is that ice-rich comets, the coldest and most distant bodies in the solar system, contain fragments of materials that make up the terrestrial planets.

The Stardust thermal protection system worked well during the mission. NASA realized that it required new technologies for reentry as its planetary exploration ramped up in the early/mid-1990s.[55] Accordingly, it pursued "development of two new lightweight ablators, Phenolic Impregnated Carbon Ablator (PICA) and Silicone Impregnated Reusable Ceramic Ablator (SIRCA). Owing to its performance in the ~ 1 kW/cm2 heating environment and low heat shield mass, PICA enabled the Stardust Sample Return Mission. SIRCA has been used on the backshell of Mars Pathfinder and the Mars Exploration Rover missions."[56] Furthermore, as one report elaborated:

---

54. Francis Reddy, "Salvaging Science from Genesis," *Astronomy*, September 18, 2004, available online at *http://www.astronomy.com/asy/default.aspx?c=a&id=2444*, accessed October 20, 2009; Ansgar Grimberg, Heinrich Baur, Peter Bochsler, Fritz Bühler, Donald S. Burnett, Charles C. Hays, Veronika S. Heber, Amy J.G. Jurewicz, and Rainer Wieler, "Solar Wind Neon from Genesis: Implications for the Lunar Noble Gas Record," *Science* 314 (November 17, 2006): 1133–1135.

55. M.A. Covington, "Performance of a Light-Weight Ablative Thermal Protection Material for the Stardust Mission Sample Return Capsule," NASA Ames Research Center TR-20070014634 (2005).

56. Ethiraj Venkatapathy, Christine E. Szalai, Bernard Laub, Helen H. Hwang, Joseph L. Conley, James Arnold, and 90 coauthors, "Thermal Protection System Technologies for Enabling Future Sample Return Missions," White Paper to the National Research Council Decadal Primitive Bodies Sub-Panel, available online at *http://www.psi.edu/decadal/topical/EthirajVenkatapathy.pdf*, accessed October 22, 2009.

Artist's rendering of the Stardust spacecraft. The spacecraft was launched on February 7, 1999, from Cape Canaveral Air Force Station aboard a Delta II rocket. The primary goal of Stardust was to collect dust and carbon-based samples during its closest encounter with Comet Wild 2—pronounced "Vilt 2" after the name of its Swiss discoverer. NASA PIA03183.

PICA was developed by NASA Ames in the early-mid 90s, is fabricated by Fiber Materials, Inc. (FMI) and was employed as the forebody TPS on the Stardust Return Capsule. It is currently the baseline forebody TPS for the Orion Crew Exploration Vehicle (CEV) and is being fabricated as the forebody TPS for the Mars Science Laboratory (MSL), scheduled for launch in mid-2009. PICA is a low density carbon-based ablator. Under both the CEV TPS Advanced Development Program and MSL [Mars Science Laboratory], an extensive data base has been developed. The failure modes are well-understood (upper heat flux limits of ≈ 1500 W/cm2 and pressure of 1.0-1.5 atm) and validated design models have been developed. For Stardust, the PICA heat shield was fabricated as one piece. But that will not be possible for larger vehicles leading to a tiled design (such as the MSL design) that introduces significant design and fabrication complexities.[57]

PICA has been flight-qualified and used in the past, and along with investments in new thermal protection system materials, it has offered acceptable, cost-effective results.[58]

Because of the high velocity of reentry for Stardust, the vehicle's forebody thermal protection systems had to use ablative materials, even as the afterbody's blunt-cone aeroshell's heating was much less hazardous to the spacecraft. Because of the stresses of forebody heating, NASA engineers used a complex analysis to arrive at the proper materials for its ablative heat shield:

> To meet the requirements for the Stardust mission, one of a family of lightweight ceramic ablator materials developed at NASA Ames Research Center was selected for the forebody heat shield of the Stardust sample return capsule. This material, phenolic impregnated carbon ablator (PICA), consists of a commercially available low density carbon fiber matrix substrate impregnated with phenolic resin.... The Stardust program resulted in intensive material development, modeling, and testing efforts to provide a heat shield

---

57. E. Venkatapathy, B. Laub, G.J. Hartman, J.O. Arnold, M.J. Wright, and G.A. Allen, Jr., "Selection and Certification of TPS: Constraints and Considerations for Venus Missions," 6th International Planetary Probe Workshop, Atlanta, GA, June 23–27, 2008.

58. H.K. Tran, W.D. Henline, Ming-tu Hsu, D.J. Rasky, and S.R. Riccitiello, patent no. 5,536,562, July 16, 1996, and patent no. 5,672,389, September 30, 1997, "Low Density Resin Impregnated Article Having an Average Density of 0.15 to 0.40 g/cc."

NASA's Stardust sample return capsule successfully landed at the U.S. Air Force Utah Test and Training Range on January 15, 2006. The capsule contained cometary and interstellar samples gathered by the Stardust spacecraft. NASA PIA03669.

for the high convective heating conditions expected during Earth entry while under constraints of limited time and funding.[59]

As it turned out, the Stardust reentry system worked well, dissipating about 90 percent of the total energy via the bow-shock heating of the atmospheric gases. Additionally, the lessons learned from this effort fed into the next major project NASA undertook for reentry and recovery from space.

The Orion Capsule as the Next Generation Piloted Vehicle

As the United States entered the first decade of the 21st century, its piloted space flight vehicle, the Space Shuttle, had been in use for 20 years, and the technology was being superseded by more advanced concepts. Even so, NASA

---

59. M.A. Covington, J.M. Heinemann, H.E. Goldstein, Y.K. Chen, I. Terrazas-Salinas, J.A. Balboni, J. Olejniczak, and E.R. Martinez, "Performance of a Low Density Ablative Heat Shield Material," *Journal of Spacecraft and Rockets* 45 (March–April 2008): 237–242.

had not been successful in replacing the vehicle with a viable successor despite having tried to do so with NASP, the X-33, the X-34, and other projects that did not reach the hardware stage. This situation reached crisis proportions in the aftermath of the tragic Columbia accident, on February 1, 2003, when the vehicle and its crew were lost during reentry. With the intention of refocusing the space agency's human space flight efforts, on January 14, 2004, President George W. Bush announced the "Vision for Space Exploration," which was aimed at human exploration of the Moon and Mars. As stated at the time, the fundamental goal of this vision was to advance U.S. scientific, security, and economic interests through a robust space exploration program. In support of this goal, the President announced that the Nation would do the following:

- Implement a sustained and affordable human and robotic program to explore the solar system and beyond;
- Extend human presence across the solar system, starting with a human return to the Moon by the year 2020, in preparation for human exploration of Mars and other destinations;
- Develop the innovative technologies, knowledge, and infrastructures both to explore and to support decisions about the destinations for human exploration; and
- Promote international and commercial participation in exploration to further U.S. scientific, security, and economic interests.

In so doing, the President called for completion of the ISS and retirement of the Space Shuttle fleet by 2010. Resources that would have been expended in keeping the Shuttle program operating could then go toward creating the technologies necessary to return to the Moon and eventually to Mars. By 2009, however, it had become clear that this program was unsustainable, and after a review of the program by a blue-ribbon panel led by former Lockheed Martin Chief Executive Officer Norm Augustine the new administration of President Barack Obama terminated the program in favor of another approach. Accordingly, the "Vision for Space Exploration" followed the path of the aborted Space Exploration Initiative, which was announced with great fanfare in 1989 but derailed in the early 1990s.[60]

---

60. Frank Sietzen, Jr., and Keith L. Cowing, *New Moon Rising: The Making of the Bush Space Vision* (Burlington, Ontario: Apogee Books, 2004); Craig Cornelius, "Science in the National Vision for Space Exploration: Objectives and Constituencies of the 'Discovery-Driven' Paradigm," *Space Policy* 21 (February 2005): 41–48; Wendell Mendell, "The Vision for Human Spaceflight," *Space Policy* 21 (February 2005): 7–10; Thor Hogan, *Mars Wars: The Rise and Fall of the Space Exploration Initiative* (Washington, DC: NASA SP-2007-4410, 2007).

In the meantime, NASA had begun efforts to build a new spacecraft to carry out the expanded space exploration mission announced in the "Vision for Space Exploration." This was intended to allow humanity to move beyond Earth toward multiplanetary activities. Named the Constellation program, it called for the reuse of as much existing Space Shuttle technology as possible to create a new Ares I crew launch vehicle. The vehicle was to consist of a Space Shuttle solid rocket booster as a first stage and an external tank as the beginning point for a second stage. A piloted space capsule, Orion, was to sit atop this system. A second rocket, the proposed Ares V cargo launch vehicle, would provide the heavy-lift capability necessary to journey back to the Moon or to go beyond. Ares I was intended to carry a crew of up to six astronauts to low-Earth orbit in the Orion crew exploration vehicle, with the capability for expanding its use to send four astronauts to the Moon. Ares V was intended to serve as the Agency's primary vehicle for delivery of large-scale hardware and cargo into space.

The Orion spacecraft, which began development as the Crew Exploration Vehicle (CEV), was only one part of the Constellation program's fleet of launchers and other vehicles designed to take humans out of Earth's orbit. According to NASA specifications:

> Orion borrows its shape from space capsules of the past, but takes advantage of the latest technology in computers, electronics, life support, propulsion and heat protection systems. The capsule's conical shape is the safest and most reliable for re-entering the Earth's atmosphere, especially at the velocities required for a direct return from the moon.
>
> Orion will be 16.5 feet in diameter and have a mass of about 25 tons. Inside, it will have more than 2.5 times the volume of an Apollo capsule. The spacecraft will return humans to the moon to stay for long periods as a testing ground for the longer journey to Mars.[61]

Conceived by NASA and built by Lockheed Martin, the Orion was originally intended to be all things to all people. It could fly to the ISS but just as easily travel to and from the Moon. Over time, NASA had to back away from that requirement in favor of more modest Earth-orbital capabilities, with the potential for a hardier translunar version of Orion in the future.[62]

---

61. NASA Press Release, "NASA Names New Crew Exploration Vehicle Orion," August 22, 2006, NASA Historical Reference Collection.

62. NASA Press Release, "NASA Names Orion Contractor," August 31, 2006, NASA Historical Reference Collection.

The Ares V (left) and Ares I (right) were part of an ambitious plan to replace the Space Shuttle with Shuttle-derived hardware that would carry out the "Vision for Space Exploration." Part of the Constellation program, announced in 2005, it was canceled in 2010 in favor of a commercially derived piloted space launcher. NASA.

The thermal protection system for Orion went through several considerations until finally, in September 2006, NASA awarded a contract to Boeing for its development. As Bill Christensen wrote in Space.com at the time:

> NASA Ames Research Center has awarded the $14 million contract for the development of a phenolic impregnated carbon ablator (PICA) heat shield. The PICA shield was first used on the Stardust interplanetary spacecraft launched February 7, 1999 to study the composition of the Wild 2 comet. It successfully reentered Earth's atmosphere January 15, 2006. The capsule was traveling at 28,900 miles per hour (46,510 kilometer per second), the fastest reentry speed ever achieved by a man-made object....
>
> The ability to survive high speed reentry is an important consideration for the Orion spacecraft. Orion is intended to perform lunar-direct returns, which result in considerably higher speeds; the spacecraft will need to withstand about five times more heat than experienced by spacecraft returning from the International Space Station.
>
> The best protection against high heat flux is an ablative heat shield. The extreme heat of reentry causes the material to pyrolize—the chemical decomposition of a material by heating in the absence of oxygen. As the PICA chars, melts and sublimates, it creates a cool boundary layer through blowing, protecting the spacecraft.[63]

To determine how best to proceed with development of the new heat shield, NASA undertook study over a 3-year period with Boeing and other entities aimed at ensuring success; its Orion Thermal Protection System Advanced Development Project considered eight different candidate materials.

The two final candidates for the Orion thermal protection system were Avcoat and phenolic-impregnated carbon ablator, both of which had proven track records in space reentry and recovery operations. NASA announced on April 7, 2009:

> Avcoat was used for the Apollo capsule heat shield and on select regions of the space shuttle orbiter in its earliest flights. It was put

---

63. Bill Christensen, "Boeing's Thermal Protection System for Orion Spacecraft," September 27, 2006, *Space.com*, available online at *http://www.space.com/businesstechnology/060927_orion_tps_technovelgy.html*, accessed October 22, 2009; Boeing News Release, "Boeing to Develop Advanced Thermal Protection System for Orion Spacecraft," September 20, 2006, available online at *http://www.boeing.com/ids/network_space/news/2006/q3/060920e_nr.html*, accessed October 22, 2009.

The Orion Crew Exploration Vehicle undergoes a test in 2010. This basic vehicle survived the cancellation of the Constellation Program and morphed into the Multi-Purpose Crew Vehicle, which was still called Orion. NASA JSC-2010E042235.

back into production for the study. It is made of silica fibers with an epoxy-novalic resin filled in a fiberglass-phenolic honeycomb and is manufactured directly onto the heat shield substructure and attached as a unit to the crew module during spacecraft assembly. PICA, which is manufactured in blocks and attached to the vehicle after fabrication, was used on Stardust, NASA's first robotic space mission dedicated solely to exploring a comet, and the first sample return mission since Apollo.

"NASA made a significant technology development effort, conducted thousands of tests, and tapped into the facilities, talents and resources across the agency to understand how these materials would perform on Orion's five-meter wide heat shield," said James Reuther, the project manager of the study at NASA's Ames Research Center at Moffett Field, Calif. "We manufactured full-scale demonstrations to prove they could be efficiently and reliably produced for Orion."

Ames led the study in cooperation with experts from across the agency. Engineers performed rigorous thermal, structural and environmental testing on both candidate materials. The team then compared the materials based on mass, thermal and structural performance, life cycle costs, manufacturability, reliability and certification challenges. NASA, working with Orion prime contractor Lockheed Martin, recommended Avcoat as the more robust, reliable and mature system.

"The biggest challenge with Avcoat has been reviving the technology to manufacture the material such that its performance is similar to what was demonstrated during the Apollo missions," said John Kowal, Orion's thermal protection system manager at Johnson. "Once that had been accomplished, the system evaluations clearly indicated that Avcoat was the preferred system."

Even with this work, NASA was unsure what direction to go in its decision-making process. "While Avcoat was selected as the better of the two candidates," NASA engineers said, "more research is needed to integrate it completely into Orion's design."[64]

Another problem quickly emerged once NASA began concentrating on Avcoat as a possible Orion thermal protection system material. As a proprietary material, the recipe for Avcoat was hard to track down; and since Avcoat had not been made for some 40 years, those familiar with the formula were few and far between. At the same time, the Orion spacecraft would have to withstand approximately five times greater reentry heating than the Apollo missions endured. So even if Avcoat could be duplicated, it had to be stronger and thicker. "We can handle the initial operating system of (reentry from) low-Earth orbit," said James Reuther, head of Orion's Thermal Protection System Advanced Development Project for NASA's Ames Research Center. "[O]n the lunar side it's a much greater challenge. We need a single heat-shield material for the lunar environment and reentry.... We're at greater risk there building a single system for both from scratch."[65]

To help with the Avcoat research, NASA came to the Smithsonian Institution's National Air and Space Museum to analyze parts of the heat shield from the Apollo spacecraft that was flown 40 years ago. " 'We started working together at the end of June [2009] to track down any Apollo-era heat shields that they had in storage,' said Elizabeth (Betsy) Pugel of the Detector Systems Branch at NASA Goddard. 'We located one and opened it. It was like a nerd Christmas for us!' " They examined the material, the connections to the spacecraft, and

64. NASA News Release 09-080, "NASA Selects Material for Orion Spacecraft Heat Shield" (April 7, 2009), NASA Historical Reference Collection.

65. Stefanie Olsen, "NASA to Test-Fly Orion Spacecraft Next Fall," August 1, 2007, *CNET News*, available online at *http:www.news.cnet.com/NASA-to-test-fly-Orion-spacecraft-next-fall/2100-11397_3-6200278.html*, accessed October 22, 2009.

the properties of the Avcoat.[66] The challenges were significant, even with what was learned through examining the Apollo heat shields. The Orion capsule was expected to withstand as much as 5,000 °F during a return to Earth from the lunar mission, and the experience of the Space Shuttle was only about 2,300 °F.[67]

Despite the challenges, work continued apace, and virtually every NASA Center had some role in it. In August 2008, NASA and Lockheed Martin engineers were successful in bonding the backshell to the demonstration spacecraft and began testing the PICA ablator to determine compression loads, vibration parameters, and reaction to acoustical patterns.[68] NASA engineers at Langley Research Center added their expertise and undertook tests on PICA and Avcoat to discover the thermal and structural behavior of the ablators attached to a honeycomb backing structure.[69] Meanwhile, NASA Glenn Research Center engineers developed especially strong bonding to affix the ablator to the spacecraft.[70] Among many other efforts, a team from NASA's Johnson Space Center, Ames Research Center, and Engineering Research Consultants Inc. undertook research to "improve the Orion (new launch vehicle) vehicle's thermal protection system (TPS) properties"[71] by adding new elements to the Avcoat formula, perhaps even carbon nanotubes. This report noted:

> Small additions of carbon nanotubes to the phenolic polymer have potential to greatly improve char strength. Char is formed as a thin layer as heat from entry pyrolizes PICA's leading surface. Aerodynamic heat and pressure from lunar and martian returns

---

66. "Apollo Heat Shield Uncrated After 35 Years, Will Help with Orion CEV," June 21, 2009, available online at *http://galaxywire.net/tag/thermal-protection-system/*, accessed October 22, 2009.

67. "Keeping it Cool," January 31, 2008, available online at *http://www.nasa.gov/mission_pages/constellation/orion/heatshield_prt.htm*, accessed October 22, 2009.

68. "Orion Crew Exploration Vehicle Weekly Accomplishments," August 15, 2008, NASA Historical Reference Collection.

69. "To the Extreme: NASA Tests Heat Shield Materials," February 3, 2009, available online at *http://www.nasa.gov/mission_pages/constellation/orion/orion-tps_prt.htm*, accessed October 22, 2009.

70. "Awards Received from CEV Thermal Protection System Advanced Development Project," September 19, 2009, available online at *http://rt.grc.nasa.gov/2009/09/19/awards-received-from-cev-thermal-protection-system-advanced-development-project/*, accessed October 22, 2009.

71. Michael Waid, Scott Coughlin, Padraig Moloney, Pasha Nikolaev, Edward Sosa, Olga Gorelik, Peter Boul, Mike Fowler, Sivaram Arepalli, Leonard Yowell, Jim Arnold, Mairead Stackpoole, Wendy Fan, and Brett Cruden, "Thermal Radiation Impact Protection System," *Engineering*, pp. 80–81, available online at *http://research.jsc.nasa.gov/BiennialResearchReport/PDF/Eng-14.pdf*, accessed October 22, 2009.

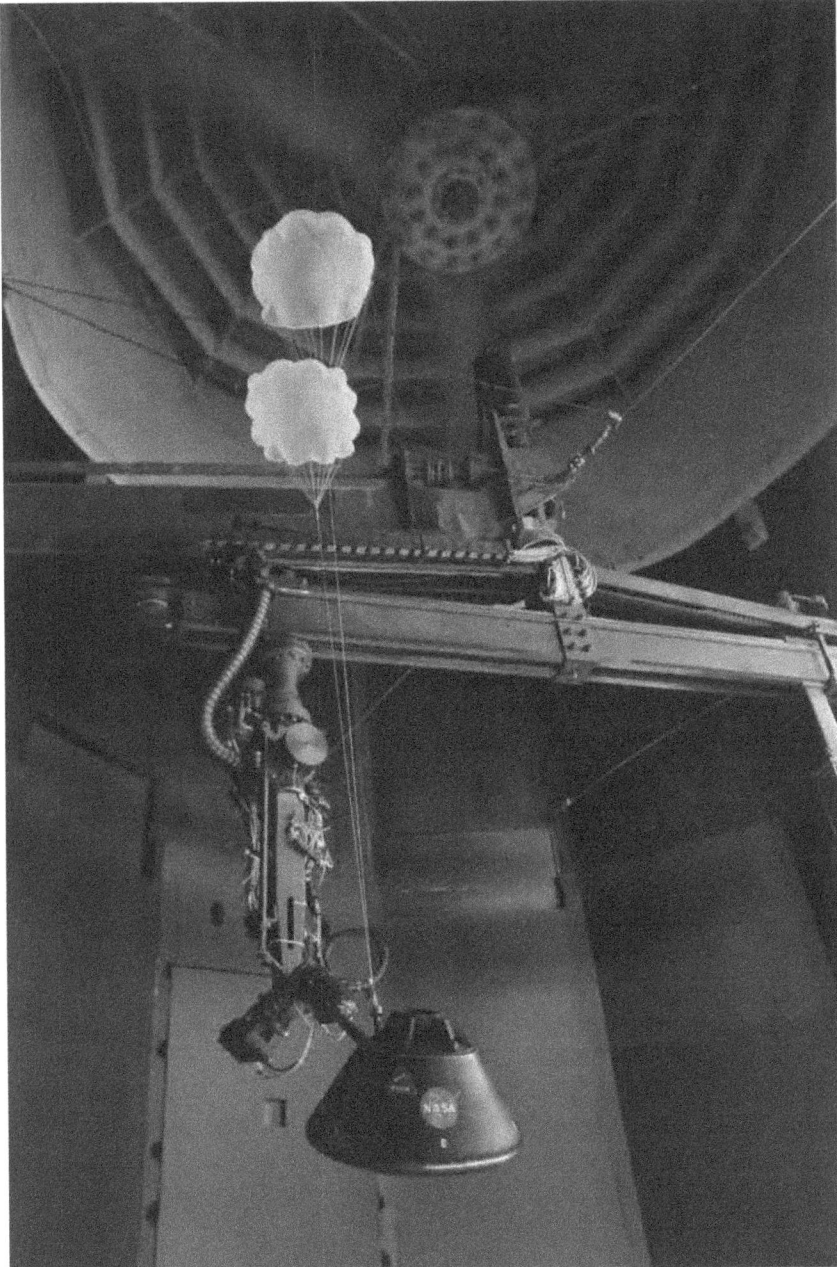

Engineers performed a series of tests in the 20-foot vertical spin tunnel in 2010 that measured the aerodynamics of a 6.25-percent model of the Orion crew module with deployed drogue parachutes. The purpose of the test was to compare how the model performed in simulated flight versus a free-flight flight test in the tunnel. NASA.

will push the ablator to its limits. If the char is too weak, pressure from hot internal gases can cause the char to spall or break off, thus exposing virgin PICA. This causes uneven heating and disturbances to the boundary layer. Carbon nanotubes increase char strength by adding an additional fibrous structure to the otherwise randomly orientated pyroloisis region.[72]

Of course, this effort is far from complete and may never enter into the Orion thermal protection system. Further study is ongoing.

In the end, the Orion thermal protection system will be a passive ablator that will work much like the system used on Apollo. The higher reentry speeds and harsher flight regime for a lunar reentry have been put off a bit with concentration on the Earth orbital reentry and recovery. It will consist of the following elements:

> For the CEV CM, spacecraft protection is the TPS that includes ablative TPS on the windward (aft) side of the vehicle, reusable surface insulation for the external leeward (central and forward) TPS, and internal insulation between the pressurized structure and OML. There are a number of potential materials available for use in the CEV CM protection system and the eventual TPS materials selected will be the result of a rigorous trade study based on performance and cost. Some of these materials may include carbon-carbon, carbon-phenolic, AVCOAT, Phenolic Impregnated Carbonaceous Ablator (PICA), PhenCarb-28, Alumina Enhanced Thermal Barrier-8 (AETB–8))/TUFI, Advanced Flexible Reusable Surface Insulation (AFRSI, LI-900 or LI-2200, CRI, SLA-561S, cork, and many others.
>
> TPS mass for the present CEV CM concept is scaled from an analysis conducted for a vehicle of the same base diameter but lower sidewall angle and higher mass at Entry Interface (EI). A 5.5-m, 28-deg sidewall concept with a total mass of approximately 11,400 kg requires an aft TPS mass of 630 kg and forward TPS mass of 180 kg. The assumed TPS materials for this analysis were PICA for the aft side and a combination of LI-2200, LI-900, AFRSI, and Flexible Reusable Surface Insulation (FRSI) at equal thicknesses for the central and forward side. The maximum heating rate for the TPS is driven by ballistic entry trajectories at lunar return speeds (11 km/s), and TPS thickness is sized by the total integrated

---

72. Ibid.

heat load of a skip-entry trajectory. For the lighter 5.5-m, 32.5-deg CM, the 630-kg aft TPS mass from the larger, heavier concept has been retained to provide additional margin, while the central and forward TPS mass has been scaled based on the lower surface area. The current CEV CM mass, including external TPS, is 9,301 kg at atmospheric EI for the nominal lunar mission.[73]

It remains to be seen how this will work out, but the Orion program has "invested considerable resources in developing analytical models for PICA and AVCOAT, material property measurements that is essential to the design of the heat-shield, in arc-jet testing, in understanding the differences between different arc jet facilities…and in integration of and manufacturing [the] heat shield as a system." Since 2005, significant strides have been made in understanding the problem and working toward its resolution. In the process, researchers have advanced the state of the art in thermal protection system technology, even as Orion engineers understood that much more remained to be done.[74]

One other issue affecting the recovery from space was the development of a new parachute system that would carry Orion home. These relatively simple systems have flown reliably in all but a handful of cases since the beginning of the space age, although there was a problem with the Genesis parachute. As envisioned, the CEV Parachute Assembly System (CPAS) would rely on proven methods of past experience:

> The CPAS will be deployed in subsonic controlled flight with the angle of attack and sideslip within 40 degrees of trim, the vehicle pitch and yaw rates less than 40 deg/sec and the vehicle roll rate less than 80 deg/sec. The CPAS will be initiated at altitudes ranging from 4,000 to 40,000 ft AGL [above ground level]. The vehicle weight at drogue deploy is assumed to be 17,176 lbs. The drogues will be deployed at dynamic pressures ranging from 19 to 115 psf and be required to stabilize and decelerate the vehicle prior to release and deployment of the mains. The vehicle weight at touchdown is assumed to be 14,400 lbs. The CPAS shall nominally deliver no greater than 26 ft/sec rate of descent and no greater than 33 ft/sec with one main failed using an air density of 0.00182526 (representing a three sigma dispersed hot day

73. NASA, "NASA's Exploration Systems Architecture Study," NASA TM-2005-214062 (November 2005), pp. 229–230.

74. Ethiraj Venkatapathy and James Reuther, "NASA Crew Exploration Vehicle, Thermal Protection System, Lessons Learned," 6th International Planetary Probe Workshop, Atlanta, GA, June 26, 2008.

at White Sands Missile Range). The CPAS shall stabilize the vehicle to within +/- 5 degrees of the desired hang angle, recognizing that the hang angle for land and water landings will likely be different.[75]

The configuration was to be similar to Apollo's, with dual drogue chutes, both sufficient to deploy the main parachutes. In addition, there would be three main parachute systems, any two of which could land the spacecraft safely. It would also employ pilot parachutes to deploy each main parachute individually; this was especially attractive because of the bell-shaped geometry of the spacecraft—like Apollo—which used a similar system.[76]

A New President Changes the Paradigm

Despite this success, by the end of the administration of President George W. Bush it had become clear that the Constellation program was not progressing as well as intended. A 2008 report summarized attitudes toward the program during the last year of the Bush administration as follows:

> Congress has been debating the Vision, including its impact on the shuttle and on U.S. human access to space. Some Members wanted to terminate the shuttle earlier than 2010 because they feel it is too risky and/or that the funds should be spent on accelerating the Vision. Others want to retain the shuttle at least until a new spacecraft, the Crew Exploration Vehicle (CEV), is available to take astronauts to and from the ISS. The CEV is now planned for 2015 at the earliest, leaving a multi-year gap during which U.S. astronauts would have to rely on Russia for access to the ISS.[77]

Even before the presidential election of 2008, however, it had become highly uncertain that the Constellation program would continue. Virtually

75. Ricardo Machin, Anthony P. (Tony) Taylor, Robert Sinclair, and Paul Royall, "Developing the Parachute System for NASA's Orion—An Overview at Inception," 19th AIAA Aerodynamic Decelerator Systems Technology Conference and Seminar, Williamsburg, VA, May 21–24, 2007.
76. "NASA Tests Launch Abort Parachute System," August 18, 2008, available online at *http://www.nasa.gov/mission_pages/constellation/orion/pa_chute_test.html*, accessed October 22, 2009.
77. Carl E. Behrens, "The International Space Station and the Space Shuttle," Congressional Research Center, Order Code RL33568 (November 3, 2008).

no political will existed in the Bush administration to make it a reality, and certainly not much additional funding was forthcoming.

After President Barack Obama took office in January 2009, he set about organizing his administration. During this organization process, the President received several reports from a variety of sources, including his own transition team for NASA, stating that the Constellation program was very much over budget and behind schedule. There was a much-reported confrontation between NASA Administrator Michael D. Griffin and the head of the transition team (and future Obama appointee as NASA Deputy Administrator) Lori B. Garver at a December 2008 event at NASA Headquarters. Garver and Griffin spoke quietly about the situation but were overheard. Garver used the metaphor of purchasing a vehicle: "Mike, I don't understand what the problem is. We are just trying to look under the hood." Griffin's response was defensive. "If you are looking under the hood, then you are calling me a liar," Griffin replied. "Because it means you don't trust what I say is under the hood."[78]

Administrator Griffin publicly voiced his complaints about those expressing concern about the Constellation program, in April 2009, when delivering a speech before the National Space Club:

> I've grown impatient with the argument that Orion and Ares 1 are not perfect, and should be supplanted with other designs. I don't agree that there is a better approach for the money, but if there were, so what? Any proposed approach would need to be enormously better to justify wiping out four years worth of solid progress. Engineers do not deal with "perfect". Your viewgraphs will always be better than my hardware. A fictional space program will always be faster, better, and cheaper than a real space program.[79]

Impatience aside, a cacophony of criticism from many quarters required investigation and perhaps alteration of the Constellation program.

What was obvious among space-policy decision makers working with the new administration was that the technological and budgetary challenges of the Constellation program had exploded since its baseline in 2005 and that the Obama administration had to take action. The President appointed a blue-ribbon panel to review the program and make recommendations. The original desire

---

78. Robert Block and Mark K. Matthews, "NASA Chief Griffin Bucks Obama's Transition Team," *Orlando Sentinel*, December 11, 2008.

79. Michael D. Griffin, "A Fictional Space Program," Goddard Memorial Dinner, National Space Club, Washington, DC, April 17, 2009.

was that a new NASA Administrator would be nominated before January 20, 2009, so that this person could commission a review. When the Administrator's selection was delayed, the White House decided to go ahead with the review. In the spring of 2009, President Obama tapped longtime aerospace official Norm Augustine to head the study effort. The President's announcement stated:

> The "Review of United States Human Space Flight Plans" is to examine ongoing and planned National Aeronautics and Space Administration (NASA) development activities, as well as potential alternatives, and present options for advancing a safe, innovative, affordable, and sustainable human space flight program in the years following Space Shuttle retirement. The panel will work closely with NASA and will seek input from Congress, the White House, the public, industry, and international partners as it develops its options. It is to present its results in time to support an Administration decision on the way forward by August 2009.

The President's mandate emphasized a four-part agenda: "1) expediting a new U.S. capability to support use of the International Space Station; 2) supporting missions to the Moon and other destinations beyond low-Earth orbit; 3) stimulating commercial space flight capabilities; and 4) fitting within the current budget profile for NASA exploration activities."[80]

Augustine's panel's report was submitted to the White House in the fall of 2009. It validated the growing concerns that the Constellation program would not be sustainable because of the pressures of budget, technology, and time. Specifically, Augustine's panel concluded in its summary report:

- Human exploration beyond low-Earth orbit is not viable under the FY 2010 budget guideline.
- Meaningful human exploration is possible under a less constrained budget, ramping to approximately $3 billion per year above the FY 2010 guidance in total resources.
- Funding at the increased level would allow either an exploration program to explore Moon First or one that follows a Flexible Path of exploration. Either could produce results in a reasonable timeframe.[81]

---

80. Office of Science and Technology Policy, Executive Office of the President, "U.S. Announces Review of Human Space Flight Plans," May 9, 2009, NASA Historical Reference Collection, NASA Headquarters, Washington, DC.

81. Augustine Panel, "Summary Report of the Review of U.S. Human Space Flight Plans Committee," September 8, 2009, available online at *http://www.nasa.gov/pdf/384767main_SUMMARY%20 REPORT%20-%20FINAL.pdf*, accessed June 9, 2011.

The final report offered more detail and went into a sustained discussion of something hinted at in the summary: the possibilities for commercial crew and cargo support for the International Space Station.

A key discussion in the final report involved the harnessing of private-sector, especially entrepreneurial, firms in supporting the International Space Station, instead of relying on the Constellation program. The report noted that a major gap would occur between the retirement of the Space Shuttle and the first use of any Constellation hardware: "Under current conditions, the gap in U.S. ability to launch astronauts into space will stretch to at least seven years. The Committee did not identify any credible approach employing new capabilities that could shorten the gap to less than six years." This would be true even with increased funding for NASA's program. The panel noted that a $3 billion-per-year increase for fiscal years 2010 to 2014 could return the Constellation program to health. But there was another option. "As we move from the complex, reusable Shuttle back to a simpler, smaller capsule, it is appropriate to consider turning this transport service over to the commercial sector," the panel concluded. "This approach is not without technical and programmatic risks, but it creates the possibility of lower operating costs for the system and potentially accelerates the availability of U.S. access to low-Earth orbit by about a year, to 2016. If this option is chosen, the Committee suggests establishing a new competition for this service, in which both large and small companies could participate."[82]

The response to this report from the space community was immediate. Some administration officials urged that the President cancel Constellation. Edward Crawley, a Massachusetts Institute of Technology professor and a member of the Augustine panel, remarked that Ares I was suffering from technical issues that could only be overcome with more money and time. "It was a wise choice at the time,"[83] said Crawley, when asked about originating the program in 2005. "But times have changed," he added. "The budgetary environment is much more tight, and the understanding of the cost and schedule to develop the Ares I has matured." Others were supportive of continuing Constellation. Constellation Program Manager Jeff Hanley defended the program and argued that the panel did not "take into account the improvements we have made in our schedule quality and risk posture through deployment of reserves and the reduction of program content."[84]

82. Augustine Panel, "Seeking a Human Spaceflight Program Worthy of a Great Nation," October 23, 2009, available online at *http://www.nasa.gov/pdf/396093main_HSF_Cmte_FinalReport.pdf*, accessed June 9, 2011.

83. Amy Klamper, "NASA in Limbo as Augustine Panel Issues Final Report," *Space News*, October 23, 2009, available online at *http://www.spacenews.com/civil/091023-nasa-augustine-panel-final-report.html*, accessed June 9, 2011.

84. Ibid.

In the midst of this turmoil, NASA made a test flight of the Ares I-X on October 28, 2009. There was a 2-minute powered flight segment, and the entire mission lasted about 6 minutes from launch until splashdown of the rocket's booster stage nearly 150 miles downrange. The rocket reached nearly 3 g's and a speed of Mach 4.76 (not quite hypersonic speed). The dummy upper stage returned to the recovery area under parachutes at a suborbital altitude of 150,000 feet after the separation of its first stage, a four-segment solid-rocket booster. As this was not a test of the reentry and recovery system, the simulated upper stage, Orion crew module, and launch-abort system were not recovered. The results of this test were mixed; some criticized that the first stage was a surplus (shelf-life expired) four-segment Shuttle solid-rocket booster and not the proposed five-segment Ares first stage.[85]

Based on these responses, President Obama proposed on February 1, 2010, with more details added in a Presidential speech on April 15, a radical new path for future U.S. human space flight efforts. Central to this was the termination of the Constellation program as a single entity, the continuation of certain technology developments such as the Orion space capsule, the continuation of operations on the International Space Station until at least 2020, and the fostering of private-sector solutions to support operations in low-Earth orbit. Since this declaration on February 1, 2010, numerous high-profile space flight advocates weighed in on both sides of the debate. In April 2010, Apollo astronauts Neil Armstrong, Gene Cernan, and James A. Lovell, Jr., famously sent the President a letter warning that the proposed changes to human space flight "destines our nation to become one of second- or even third-rate stature." Proponents of the plan, among them Apollo astronaut Buzz Aldrin, counter that the President's approach will return NASA to its roots as an R&D organization while private firms operate space systems. Turning low-Earth orbit over to commercial entities could then empower NASA to focus on deep space exploration, perhaps eventually sending humans to Mars or elsewhere.[86]

The debate has largely been over maintaining a traditional approach to human space flight with NASA dominating the effort, owning the vehicles, and operating them through contractors. That was the method whereby the United States went to the Moon; it has proven successful for over 50 years of

---

85. NASA Constellation Program, "Ares I-X Flight Test," available online at *http://www.nasa.gov/ mission_pages/constellation/ares/flighttests/aresIx/index.html*, accessed June 9, 2011.

86. Michael Sheridan, "Neil Armstrong, James Lovell Call Obama's Plans For Space Exploration, NASA, 'Misguided'," *New York Daily News*, April 14, 2010; Stephanie Condon, "Neil Armstrong Vs. Buzz Aldrin Over Obama's Space Plans," *CBS News*, available online at *http://www.cbsnews. com/8301-503544_162-20002451-503544.html*, accessed June 9, 2011.

NASA's Ares I-X demonstration vehicle on LC-39B at the Kennedy Space Center, FL, on October 26, 2009. The flight test of Ares I-X, offered an opportunity to test and prove flight characteristics, hardware, facilities, and ground operations associated with the Ares I. NASA JSC2009E225491.

human space exploration. Then there are those from the "new space" world that emphasize allowing private-sector firms to seize the initiative and pursue entrepreneurial approaches to human space flight. Advocates of the more traditional approach believe that the other side will sacrifice safety; those who support the entrepreneurial approach say that those who support a more robust Government-led program are advocating large, over-budget, underachieving space efforts. It remained unclear how much, if any, of this new initiative that the U.S. Congress will approve; in 2011, there still was no resolution. Meanwhile, NASA began moving forward with the "program of record" and at the same time planning for programs to replace it.

NASA's Commercial Crew and Cargo Program

During the next year, NASA pursued efforts to replace access to low-Earth orbit, previously provided by the Space Shuttle, through a multiphase space technology development program known as Commercial Crew Development (CCDev). Intended to stimulate development of privately operated crew vehicles to low-Earth orbit, CCDev's first phase offered a token sum of $50 million during 2010 to five American companies for R&D into human space flight concepts and technologies in the private sector. In its second phase, with contracts of $269 million awarded to four firms in April 2011, the objective was to move toward the establishment of one or more orbital space flight capabilities on which NASA could purchase cargo and eventually transport crews into space.[87] The concepts, which ranged from lifting body to capsule spacecrafts, were awarded to Blue Origin, Sierra Nevada Corp., Space Exploration Technologies (SpaceX), and Boeing. As announced by Ed Mango, NASA's commercial crew program manager: "The next American-flagged vehicle to carry our astronauts into space is going to be a U.S. commercial provider. The partnerships NASA is forming with industry will support the development of multiple American systems capable of providing future access to low-Earth orbit."[88]

The spacecraft nearest to being ready for flight might well be the SpaceX Dragon capsule launched atop the Falcon 9 rocket. With its successful suborbital

---

87. NASA Press Release 10-277, "NASA Seeks More Proposals On Commercial Crew Development," October 25, 2010, available online at *http://www.nasa.gov/home/hqnews/2010/oct/HQ_10-277_CCDev.html*, accessed June 9, 2011.

88. Chris Bergin, "Four Companies Win Big Money via NASA's CCDEV-2 Awards," April 18, 2011, available online at *http://www.nasaspaceflight.com/2011/04/four-companies-win-nasas-ccdev-2-awards/*, accessed June 9, 2011.

test flight on December 8, 2010, the SpaceX entry into this competition appeared destined for an early operational date. The reentry and recovery technology on Dragon were virtually identical to those used by the Orion vehicle, especially the PICA heat-shield technology pioneered at NASA. Its three-canopy parachute recovery system was also modeled on those used in earlier programs.[89]

The second awardee, garnering the largest award of $92.3 million, was Boeing. Its CST-100 crew capsule had been pursued as a support vehicle for the ISS. Very close in design to the Orion spacecraft of the Constellation program (but without the deep space capability), Boeing's vehicle was intended to house crews as large as seven and could be attached to ISS for more than half a year before requiring relief. For reentry and recovery, Boeing intended to use three parachutes for landing and its Lightweight Ablator for the heat shield. In every case, Boeing has emphasized its long history in working on these technologies and the legacy reentry and recovery systems that the company intended to use in the CST-100. Boeing officials have projected flight for this vehicle in 2015.[90]

Another recipient of the CCDev award, but receiving only $22 million, was Blue Origin, a startup entrepreneurial firm that proposed developing a biconic orbital capsule launched atop an Atlas V. Information about this project is limited, and Blue Origin is reticent to speak in public about its activities. Dan Rasky of NASA Ames remarked: "I joke with people that if you want to see what a billionaire's clubhouse looks like, go visit Blue Origin."[91] Little information is available as yet about the Blue Origin capsule's thermal protection and recovery systems. In Phase 2A of the effort, the company would develop "a pressurized 'Crew Return Vehicle' capsule with TPS, solid escape system and a landing mechanism (whether that's VTVL rocket or parachutes or both). Partially funded by CCDev. Give the solid escape system sufficient

---

89. SpaceX, "Dragon Lab Fact Sheet," available online at *http://www.spacex.com/downloads/dragonlab-datasheet.pdf*, accessed June 9, 2011.

90. "Commercial Human Spaceflight Plan Unveiled," *Aviation Week* (July 20, 2010); Denise Chow, "New Boeing Spaceship Targets Commercial Missions," *Space.com*, June 25, 2010, available online at *http://www.space.com/8655-boeing-spaceship-targets-commercial-missions.html*, accessed June 10, 2011; "NASA CCdev Space Act Agreement with Boeing," available online at *http://www.nasa.gov/centers/johnson/pdf/444144main_nnj10ta03s_boeing_saa.pdf*, accessed June 10, 2011; Keith Reiley, the Boeing Company, "Boeing CST-100 Commercial Crew Transportation System," February 2011, available online at *http://www.aiaa.org/pdf/industry/presentations/keith_reiley.pdf*, accessed June 10, 2011.

91. "Blue Origin Proposes Orbital Vehicle," February 18, 2010, *NewSpace Journal*, available online at *http://www.newspacejournal.com/2010/02/18/blue-origin-proposes-orbital-vehicle/*, accessed June 10, 2011.

capability abort at liftoff and sufficient TPS return from Mach 10 trajectories (and structure to beef up TPS to orbital velocities)."[92]

Finally, the CCDev program awarded an $80 million contract to the Sierra Nevada Corp. to build a commercial Space Transportation System based on NASA HL-20 and launched on an Atlas V. This Dream Chaser lifting body spacecraft would utilize Virgin Galactic's carrier aircraft as a platform for atmospheric drop tests as early as 2012. As a lifting body, this vehicle is intended "to land on a runway and be reusable. Its carbon composite airframe looks large but because of the composite structure it weighs only 27,100 pounds. For all of the spaceplane's sleekness, as one commentator noted, it could handle only 1500K cargo and only 2 crew. This is a good looking vehicle, but it is not a mini Shuttle. It is a minuscule Shuttle." [93] Its thermal protection system, at present, is unclear but will probably use a combination of RCC panels, ceramic tiles, and composites.

The Air Force and the X-37B

As NASA engaged in this highly public effort to develop a new human space flight capability to succeed the Space Shuttle and support the ISS, the U.S. Air Force quietly undertook its own spaceplane development effort: the X-37B. This program had originated as a NASA effort in August 1998, when a research announcement solicited proposals for "Future-X"—a flight demonstrator designed to validate emerging hypersonic technologies leading toward reductions in the cost of space access. The announcement specifically called for the development of propellant tanks, thermal protection systems, avionics, and structures—especially thorny technology issues that required more capable systems if efficiencies in space access were to be significantly advanced. NASA contracted with the Boeing Company of Seal Beach, CA, in July 1999, for a 4-year cooperative agreement to develop what became known as the X-37 advanced-flight demonstrator. Not a large effort, the 4-year cooperative agreement eventually amounted to something over $500 million, with a 50/50 cost-sharing ratio between Government and industry.[94]

---

92. Online comments, March 3, 2011, available online at *http://forum.nasaspaceflight.com/index. php?topic=10685.155;wap2*, accessed June 10, 2011.

93. "Interesting NASA Briefing on Various Commercial Crew Concepts," *Nasawatch.com*, June 2, 2011, available online at *http://nasawatch.com/archives/2011/06/interesting-nas.html*, accessed June 10, 2011.

94. NASA Facts, "X-37 Technology Demonstrator: Blazing the Trail for the Next Generation of Space Transportation Systems," FS-2003-09-121-MSFC (September 2003).

The X-37 represented the third in a succession of efforts to develop and test technologies required to build a spaceplane at NASA (after the X-33 and the X-34). This X-37 was intended to build on those earlier efforts and eventually reach orbital flight regimes. To be launched shrouded aboard an expendable launcher, the X-37 was a 120-percent scale derivative of the Air Force's unpowered X-40, developed in the Air Force Space Maneuver Vehicle program, which flew seven successful unpowered approach and landing tests at Dryden Flight Test Center in 2001. The X-37 technology demonstrator effort was also intended to feed into NASA's Orbital Space Plane program. Had the X-37 flown, NASA at one point intended to deploy it from the Space Shuttle, but in the context of changing plans for Shuttle replacement the program fell by the wayside.[95]

As DARPA was wrapping up airdrop tests on the Approach and Landing Test Vehicle (ALTV), the Air Force's Rapid Capabilities Office (AFRCO) began contracting for the X-37 program. The X-37, now the X-37B, had its first drop test on April 7, 2006, at Edwards AFB, and flew an orbital test mission on April 22, 2010. Launched inside a shroud atop an Atlas V rocket, the X-37B orbited Earth until December 3, 2010, at which point it became publicly known because of its success as a robotically landed spaceplane. This was the first test of the vehicle's heat shield and hypersonic aerodynamic-handling characteristics. The launch of a second X-37B on March 5, 2011, and its successful operation in orbit suggest that this program shows great promise to, in the words of the Secretary of the Air Force, complete "risk reduction, experimentation, and operational concept development for reusable space vehicle technologies, in support of long-term developmental space objectives."[96]

Developing an integrated thermal protection system for the X-37B was one of the key objectives of the program. This effort included the following:

- Flight demonstration of an integrated thermal protection system.
- Level 1 Requirement: provide maturation and validation of thermal protection system within the confines of the X-37 reentry heating environments, which include:
    - Leading edge >2,950 °F.
    - Acreage thermal protection system at 2,400 °F.
    - High-temp gap fillers and seals to support 2,950 °F leading edge and 2,400 °F acreage.

---

95. Ibid.

96. Leonard David, "U.S. Air Force Pushes For Orbital Test Vehicle," *Space.com*, November 17, 2006, available online at *http://www.space.com/3124-air-force-pushes-orbital-test-vehicle.html*, accessed June 10, 2011.

The X-37B Orbital Test Vehicle waits in the encapsulation cell at the Astrotech facility in Titusville, FL, on April 5, 2010. Half of the Atlas V 5-meter fairing is visible in the background. U.S. Air Force.

– Durability/Re-Usability of thermal protection system better than existing Shuttle.
– Thermal protection system components 10 times more durable than current tile in windward high-temperature environments.
– New thermal protection system enables adverse weather flight conditions.[97]

The X-37B used a hot-structure control surface developed by Boeing–Huntington Beach/Seal Beach with major support from GE Power Systems Composites, and with assistance from the Air Force Research Laboratory and NASA Langley Research Center. The effort required the design, analysis, manufacture, and testing of two flaperons and ruddervators for the X-37 reentry vehicle. From there, research and testing of full-acreage thermal protection systems followed, and they were proven in the first orbital flight. The thermal protection system used silica tiles impregnated with toughened unipiece fibrous insulation. This technology has been well-tested on the Space Shuttle since 1994, when it was first used on STS-59. These tiles were hardier than earlier

---

97. Peter J. Erbland, "Current and Near-Term RLV/Hypersonic Vehicle Programs," RTO-EN-AVT-116, AFRL Air Vehicles Directorate, available online at *http://ftp.rta.nato.int/public//PubFullText/RTO/EN/RTO-EN-AVT-116///EN-AVT-116-02.pdf*, accessed June 10, 2011.

Shuttle thermal protection systems elements and were used extensively on the X-37's underside. These TUFI tiles are less susceptible to impacts since the surface material permeates the underlying insulation.

A public report on the X-37 program included the following description of the new developments in its thermal protection system:

> The X-37's most notable thermal advance is on the wing lead-ing edge. On the shuttle, that vulnerable area was covered with reinforced carbon-carbon; the X-37 uses a different material, called TUFROC, for Toughened Uni-piece Fibrous Reinforced Oxidation-Resistant Composite. TUFROC (pronounced "tough rock") was developed at NASA's Ames Research Center in California by a group led by David Stewart, who has worked on thermal protection systems since the shuttle program.
>
> Stewart explains that during reentry, heat is generated not just by friction of the vehicle against the atmosphere, but also by atoms on the surface recombining. In the shuttle's case, the carbon-carbon oxidizes. As the name implies, the new material resists oxidative damage. The surface of the shuttle's tiles heats up very fast because the insulator's high-density coating is very thin. TUFROC's surface material is thicker, and therefore takes longer to heat up. And the new material will reduce weight, which will enable the spaceplane to carry more payload.[98]

Compared to the X-20 Dyna-Soar by project officials, the X-37 appears to hold promise as a future spaceplane that might satisfy both robotic and perhaps after modification, human missions.

Summary

Quite a lot of activity in the last decade has affected the course of reentry and recovery from space. After giving up on the winged, reusable Space Shuttle and its unique thermal protection system and runway landing capability, on January 14, 2004, President George W. Bush pressed the reset button by man-dating that NASA focus on a new Moon/Mars exploration agenda using a

---

98. Michael Klesius, "Space Shuttle Jr.," *Air & Space/Smithsonian*, January 1, 2010, available online at *http://www.airspacemag.com/space-exploration/Space-Shuttle-Jr.html?c=y&page=1*, accessed June 10, 2011.

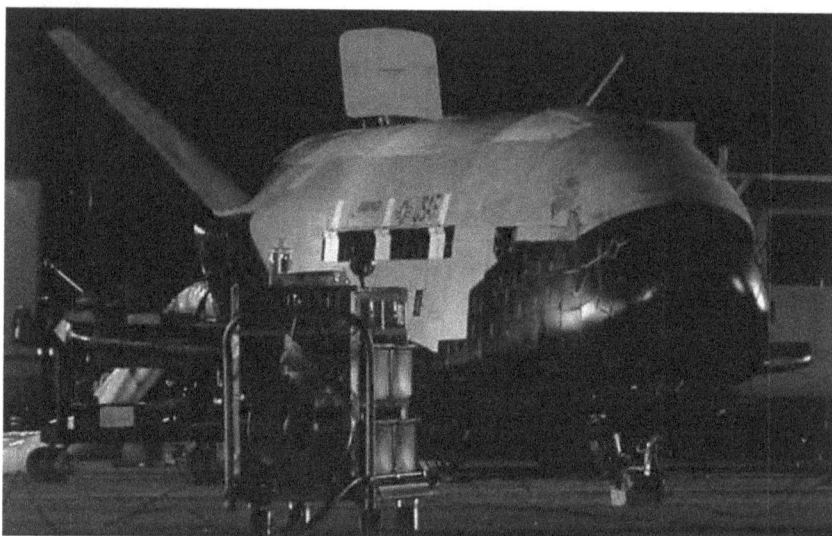

The U.S. Air Force's X-37B landed after several months in Earth orbit on December 3, 2010, at Vandenberg AFB, CA. Ironically, the runway at Vandenberg had been extensively modified to accommodate the Space Shuttle. U.S. Air Force.

capsule called the Crew Exploration Vehicle, thereby overturning any other initiative.[99] That program remained in place only until 2009, when President Barack Obama overturned it in favor of a commercial effort to replace the Space Shuttle. Although the new commercial vehicle might also be a capsule, by 2011 NASA's CCDev program was supporting yet another design for a winged spaceplane "very reminiscent of the Shuttle's design intended to ferry crews to and from the ISS."[100] Additionally, the CCDev program also supported three capsule concepts, and in every case the reentry and recovery systems

99. Leonard David, "The Next Shuttle: Capsule or Spaceplane?" *Space.com*, 21 May 2003, available online at *http://www.space.com/businesstechnology/technology/osp_debate_030521.html*, accessed April 19, 2004; Leonard David, "NASA's Orbital Space Plane Project Delayed," *Space.com*, November 26, 2003, available online at *http://www.space.com/businesstechnology/technology/space_plane_delay_031126.html*, accessed April 19, 2004; White House Press Release, "Fact Sheet: A Renewed Spirit of Discovery," January 14, 2004, available online at *http://www.whitehouse.gov/news/releases/2004/01/20040114-1.html*, accessed April 4, 2004.

100. Clay Dillow, "Jumping into the New Space Race, Orbital Sciences Unveils Mini-Shuttle Spaceplane Design," *Popular Science*, December 16, 2010, available online at *http://www.popsci.com/technology/article/2010-12/jumping-new-space-race-orbital-sciences-unveils-mini-shuttle-spaceplane-design*, accessed February 27, 2011.

were reminiscent of earlier heat shields and parachute systems. Although only the Dragon capsule had flown, albeit on a suborbital trajectory, the challenges of reentry and recovery were not major areas of concern for these programs. Instead, the programs appeared intent on using legacy technologies pioneered by NASA in some cases as far back as the Apollo program. Thus far, it appears that only the Boeing thermal protection system, developed for other systems in the recent past, pushed the envelope of knowledge about reentry. As reported in 2010: "Both capsules and space planes have their advantages, and neither has a spotless safety record. But it will be interesting to see which mode NASA eventually selects for the next generation of ISS missions."[101]

The X-37B is an entirely different type of program. Pursued as a technology demonstrator that may well lead to an operational orbital spaceplane, the Air Force has been diligent in pressing the frontiers of new technologies for thermal protection systems. By all accounts (although the details are classified for national security purposes), this system has been quite effective. As reported at the time of the first test flight:

> The success of OTV-1 on its first orbital flight, USA-212, was nothing short of spectacular, accomplishing every mission objective with a precision that stunned the aerospace world. Despite blowing out a tire on landing, which caused some minor damage to the underside of OTV-1, the spacecraft is in excellent shape and being refurbished for a second flight sometime in late 2011. The second X-37B, OTV-2, is presently being prepared for flight, with an announced liftoff scheduled for March 4, 2011, again onboard an Atlas V 501 booster. Plan on seeing a regular stream of these launches, especially if the USAF pursues the program further.[102]

With this large number of new space flight vehicles under various stages of completion, one may well be successful in providing future human space access to low-Earth orbit and the International Space Station, as well as safe reentry and recovery from space.

---

101. Ibid.

102. John D. Gresham, "X-37B Orbital Test Vehicle," *DefenseMediaNetwork*, February 28, 2011, available online at *http://www.defensemedianetwork.com/stories/x-37b-orbital-test-vehicle/*, accessed June 10, 2011.

This image from July 2008 shows the aeroshell for NASA's Mars Science Laboratory. During the journey from Earth to Mars, the aeroshell encapsulates the rover, and it shields the rover from the intense heat of friction with the upper Martian atmosphere during the initial portion of descent. The aeroshell has two main parts: the backshell, which is on top in this image and during the descent, and the heat shield, which is on the bottom. The heat shield in this image is an engineering unit used for testing. The heat shield to be used in flight will be substituted later. The heat shield has a diameter of about 15 feet. By comparison, the heat shields for the Mars Exploration Rovers Spirit and Opportunity were 8.5 feet, and the heat shields for the Apollo capsules that protected astronauts returning to Earth from the Moon were just less than 13 feet. NASA PIA11430.

CHAPTER 7

Summary

The success of atmospheric reentry missions is constrained by the design of the thermal protection systems of the aerospace vehicles involved. During reentry into Earth's atmosphere, space vehicles must operate in intense thermal-stress regimes that require an effective thermal protection system to ensure survival of the craft. Indeed, the thermal protection system may be the most important system used on a spacecraft that must enter an atmosphere and land. Next most significant, of course, is the landing system, whether it be a parachute, rocket, glide system, or Rogallo wing.

One recent study made the observation about the criticality of these systems as follows:

> For vehicles traveling at hypersonic speeds in an atmospheric environment, TPS is a single-point-failure system. TPS is essential to shield the vehicle structure and payload from the high heating loads encountered during [reentry].... Minimizing the weight and cost of TPS, while insuring the integrity of the vehicle, is the continuing challenge for the TPS community.[1]

The modern thermal protection system originated in the decade after World War II as part of the ballistic missile program. Required to ensure the reentry of nuclear warheads, the thermal protection system represented a fundamental technology for these missile systems. The ablative thermal protection system protected the vehicle through a process that lifted the hot shock-layer gas away from the vehicle, dissipated heat absorbed by the ablative material as it burned away, and left a char layer that proved remarkably effective as an insulator that blocked radiated heat from the shock layer.

---

1. Ethiraj Venkatapathy, Christine E. Szalai, Bernard Laub, Helen H. Hwang, Joseph L. Conley, James Arnold, and 90 coauthors, "White Paper to the NRC Decadal Primitive Bodies Sub-Panel: Thermal Protection System Technologies for Enabling Future Sample Return Missions" (2010), p. 2, NASA Historical Reference Collection.

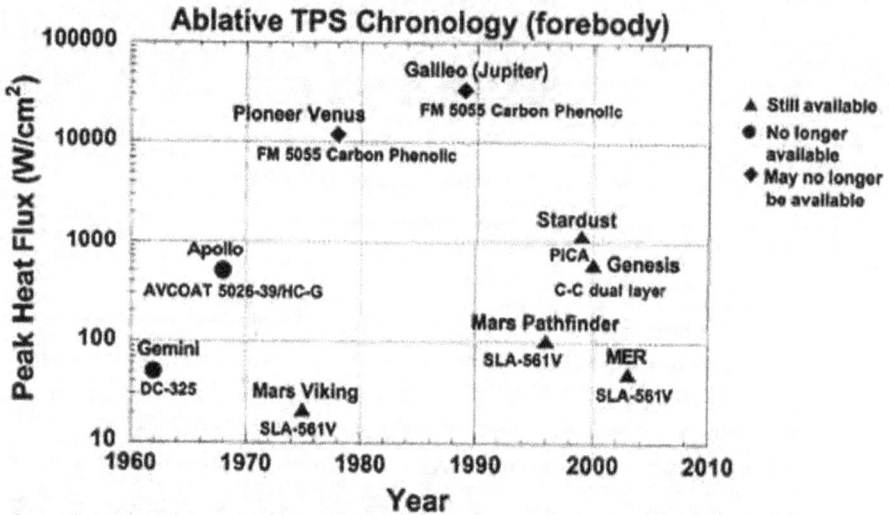

Ablative TPS Chronology (forebody)

In over 40 years, NASA entry probes have only employed a few ablative TPS materials. Half of these materials are (or are about to be) no longer available.

This ablative thermal protection system chronology demonstrates the evolution of the capability from the start of the space age to the present. NASA.

Throughout the 1960s and much of the 1970s, engineers involved in ablative thermal protection system development were quite active. The succession of programs, both human and robotic, kept them involved in making ever more capable thermal protection systems. This changed in the mid- to late 1970s, however, as R&D and testing of ablative thermal protection system materials significantly declined for both the nuclear missile programs and the robotic probes that NASA pursued. At the same time, human programs that had used ablative systems were completed, and NASA moved toward a nonablative, reusable thermal protection system for the Space Shuttle program. This transition involved significant R&D, but by the early 1980s the systems had been resolved, and only intermittent upgrades would follow. Accordingly, the ablative thermal protection system community experienced a serious decline in capability. A return to an ablative thermal protection system for robotic entry probes in the last 20 years has required a reconstitution of the knowledge base lost in the 1970s. As a result, in the 1990s NASA invested in the development of two new lightweight ablators: phenolic-impregnated carbon ablator (PICA) and silicone-impregnated reusable ceramic ablator (SIRCA). It was PICA that enabled the Stardust Sample Return Mission. Additionally, SIRCA was used on the backshell of the Mars Pathfinder and the Mars Exploration Rover missions.

These efforts reached a serious crescendo when NASA began developing the Orion crew capsule to replace the Space Shuttle. In essence, we have come full circle in the context of thermal protection systems, albeit with a spiral upward in capability and flexibility, and much of the resulting work is critically based on the efforts of earlier systems dating back to the Apollo era.[2]

A key realization coming from the cyclical process of ablative thermal protection system efforts was that despite the desire to use off-the-shelf materials, the knowledge base had atrophied so significantly that considerable effort had to be undertaken to reconstitute that knowledge base. Moreover, engineers working on thermal protection systems found that the knowledge gained in other programs was profound. For example, several engineers have concluded that without the investment in the 1990s in PICA thermal protection systems, the Mars Science Laboratory would have fallen even further behind schedule.[3] In essence, the space thermal protection system community worked most effectively when it coordinated efforts to codevelop materials and technology that could be used in multiple projects and programs. As one study concluded: "The *very* important lesson learned here is that it is wise to have at least two viable candidate TPS materials in place for mission projects because although the selected TPS sufficed for a previous project, it may not be adequate for the next, even if the entry environments are only slightly more severe."[4]

The process of moving from ablative heat shields in the 1960s to the Space Shuttle thermal protection system in the 1970s and to the various types of thermal protection systems currently available has represented something of a back-to-the-future approach to space flight. It raises important questions about the history of technology and its usually presumed progress. This history looks more circular than progressive when considering spacecraft return and recovery technology throughout the space age. It also illustrates the inexact nature of space technology and the lack of a clear line of development as the technology has looped back and forth, jumped ahead, and backtracked to earlier, proven

---

2. Bernie Laub, "Ablative Thermal Protection: An Overview," presentation at the 55th Pacific Coast Regional and Basic Science Division Fall Meeting, Oakland, CA, October 19–22, 2003.

3. Jean-Marc Bouilly, Francine Bonnefond, Ludovic Dariol, Pierre Jullien, and Frédéric Leleu, "Ablative Thermal Protection Systems for Entry in Mars Atmosphere: A Presentation of Materials Solutions and Testing Capabilities," 4th International Planetary Probe Workshop, Pasadena, CA, June 27–30, 2006.

4. Venkatapathy, Szalai, Laub, Hwang, Conley, Arnold, and 90 coauthors, "White Paper to the NRC Decadal Primitive Bodies Sub-Panel: Thermal Protection System Technologies for Enabling Future Sample Return Missions" (2010), p. 2.

Current capabilities: TPS for sample return missions

Density	Forebody Heat Shield	Supplier	Flight Qual or TRL	Potential Limit		Entry velocity, km/s		Other Potential Missions
				Heat flux, W/cm^2	Pressure, atm	<13	>13	
	FOREBODY HEAT SHIELD							
Low-Mid	PICA	FMI	Stardust	~1200	<1	●	◡	SR, CEV, Mars
	Avcoat	Textron	Apollo	~1000	<1	●	✕	Venus (aerocapture)
	ACC	LMA/C-Cat	Genesis	>2000	>1	◡	◡	MSR, CEV, Mars
	BPA	Boeing	TRL 3–4	~1000	~1	◡	✕	Venus (aerocapture)
	PhenCarb Family	ARA	TRL 5–6	(1000–4000)	>1	◡	◡	MSR, CEV, Venus, Earth
High	3DQP	Textron	DOD (TRL4)	~5000	>1	☐	☐	SR, Venus
	Heritage Carbon phenolic	Several capable, none active	Venus, Jupiter	(10,000–30,000)	>>1	☐	☐	MSR, Venus, Jupiter, Saturn, Neptune
● Fully Capable ◡ Potentially capable, qual needed ☐ Capable but heavy ✕ Not capable								

concepts.[5] The history also suggests that very little is fully defined in space technology, especially as older concepts are resurrected and brought to the fore years after their abandonment.

Ultimately, this situation suggests one firm lesson: Thermal protection system technologies for many space missions are unique to NASA, challenging, and cross cutting. Developing these systems requires specialized efforts, but they end up offering multiple uses.

Moreover, since the thermal protection system forms the external surface of any space vehicle, it must be designed for all the environments experienced by the vehicle. As one researcher concluded:

---

5. This is very effectively laid out in John Law, "Technology and Heterogeneous Engineering: The Case of Portuguese Expansion," pp. 111–134, and Donald MacKenzie, "Missile Accuracy: A Case Study in the Social Processes of Technological Change," pp. 195–222, both in Wiebe E. Bijker, Thomas P. Hughes, and Trevor J. Pinch, eds., *The Social Construction of Technological Systems: New Directions in the Sociology and History of Technology* (Cambridge, MA: The MIT Press, 1987).

The primary function of the TPS is to protect the vehicle and its contents from aerodynamic heating, so it must be sized to keep internal temperatures within acceptable limits. In addition, the TPS must maintain its structural integrity so that it provides an acceptable aerodynamic surface during all portions of flight through the atmosphere. The TPS panels must therefore be designed to withstand aerodynamic pressure and drag loads, acoustic and dynamic loading from the engines at launch, dynamic pressures that can cause panel flutter, thermal expansion mismatches between the TPS and underlying structure, and strains induced by primary vehicle loads on the underlying structure. The TPS must also have an acceptable risk of failure after low-speed impacts during launch and hypervelocity impacts from orbital debris in space. Of course, all of these functions must be accomplished while eliminating any unnecessary mass.[6]

The requirements of these systems might evolve considerably over the various design efforts of the spacecraft, and the development of an adequate thermal protection system is always challenging and time consuming. Generic solutions do not seem to work, and only with specific applications for unique situations combined with rigorous testing in all of the envisioned elements of the spacecraft's thermal protection system has the current generation of ablative heat shields emerged for use indefinitely into the future.

Some have suggested that development of a new thermal protection system material may only be accomplished through a careful balance between thermal performance and thermal structural integrity. "Regardless of whether the heat shield design is a tiled system (PICA), or a monolithic system (Avcoat), thermal-structural capabilities are critical. Detailed thermal response must be understood for the integrated system not just for acreage TPS material. Penetrations and closeouts require significant work and are difficult [to] manage…due to changing requirements."[7] For all of the success enjoyed in reconstituting an ablative heat shield capability since the 1990s, many analysts warn that future neglect of thermal protection system development—as was the case in the 1970s and 1980s—will again necessitate "an expensive, high risk, critical path approach to recover. Without the fortuitous timing of the CEV TPS ADP

---

6. M.L. Biosser, R.R. Chen, I.H. Schmidt, C.C. Poteet, and R.K. Bird, "Advanced Metallic Thermal Protection System Development," AIAA 2002-0504 (2002), p. 3.

7. Ethiraj Venkatapathy and James Reuther, "NASA Crew Exploration Vehicle, Thermal Protection System, Lessons Learned," 6th International Planetary Probe Workshop, June 26, 2008.

PICA heat shield effort, MSL would have had no TPS options to meet their… launch window."[8] Indeed, researchers seem to agree that at present, "NASA/ US [is] short of efficient, robust TPS materials for future exploration missions: high mass Mars entry, outer planets, Venus, extra-Lunar Earth return."[9]

There has been a similar circular trajectory in the history of spacecraft recovery systems. The first landing systems employed parachutes, followed by the aborted development of paraglider systems, the testing of lifting bodies, and the Space Shuttle's winged design that allowed runway landings. Finally, spacecraft landing systems have returned to the use of parachutes. For the longest time, NASA explored ways to avoid landing its astronaut missions in the sea, which requires the U.S. Navy to perform overtaxing water rescues. Furthermore, the astronauts, who were all pilots, found the method objectionable. This was eloquently communicated in 1965 when artist Wen Painter, at NASA's Flight Research Center, drew a powerful cartoon depicting two astronauts sitting in a Gemini capsule bobbing in the water while the Navy steams around in the distance, juxtaposed with a second image of a sleek lifting body landing on a runway with astronauts walking across the tarmac. The caption read: "Don't be rescued at sea; fly back in style." This cartoon captured magnificently what some saw as a key difference between space-capsule splashdowns at sea and spaceplane landings on a runway.

As soon as opportunity permitted, NASA moved to a winged spacecraft that allowed a fine degree of control over landing and ensured that splashdowns were a thing of the past. As one analyst wrote: "The 'golden age' had the astronauts flying in a spacecraft that was largely a closed system. The Shuttle, on the other hand, allows for a significant amount of control, but lacks the cachet and interest of its predecessor programs. The irony of the situation is that the Shuttle lands like an airplane in an era when test pilots are no longer the crème de la crème of astronaut candidates."[10] As the Space Shuttle was retired in 2011, NASA is wrestling anew with these landing systems. The simplest approach remains using parachutes deployed to slow the spacecraft and gently return it to Earth. With new aerodynamic systems capsules, landings might be well controlled and, whether landing on soil or water, should be quickly recovered. Orion, Dragon, and other possible future piloted spacecraft all feature this approach to landing. While considerable effort remains to be made on these

---

8. Ibid.

9. Ibid.

10. Amy Teitel, "Of Space Shuttles and Landing Systems," *Vintage Space*, November 27, 2010, available online at *http://vintagespace.wordpress.com/2010/11/27/of-space-shuttles-and-landing-systems/*, accessed June 22, 2011.

new reentry and recovery systems, this seems to be the direction of space policy for the foreseeable future.

In the context of the larger story of reentry and recovery from space, some five distinct lessons emerge that are worthy of discussion. These may be encapsulated as stated in a study of failures concerning the Space Shuttle, but they are applicable across a broad spectrum of activities:

1. **Operating Experience:** People and organizations need to learn valuable lessons from internal and external operating experience to avoid repeating mistakes and to improve operations.
2. **Mission and External Influences:** Budget and schedule pressures must not override safety considerations to prevent unsound program decisions.
3. **Normalizing Deviations:** Routine deviations from an established standard can desensitize awareness to prescribed operating requirements and allow a low-probability event to occur.
4. **Technical Inquisitiveness:** To ensure safety, managers need to encourage employees to freely communicate safety concerns and differing professional opinions.
5. **Focus on Planning and Prevention:** Safety efforts should focus more on planning and preventive actions rather than investigations and corrective actions resulting from accidents or events.[11]

In the end, the focus on thermal protection and landing systems cannot be deemphasized. As a recent study reported: "NASA's ambitious exploration vision requires TPS *innovations*. Many future missions require TPS materials and/or concepts not currently available or, in some cases, new versions of old materials. New TPS materials, ground test facilities, and improved analysis models are required and will take some time to develop[.] Advances and improved TPS capabilities will benefit an array of missions (and *enable* some)."[12]

---

11. Department of Energy Action Plan, "Lessons Learned from the Columbia Space Shuttle Accident and Davis-Besse Reactor Pressure-Vessel Head Corrosion Event" (July 2005), copy in possession of authors.

12. Bernard Laub and Ethiraj Venkatapathy, "Thermal Protection System Technology and Facility Needs for Demanding Future Planetary Missions," presentation at International Workshop on Planetary Probe Atmospheric Entry and Descent Trajectory Analysis and Science, Lisbon, Portugal, October 6–9, 2003, p. 8.

Three canopies in the sunset, Apollo returns to Earth. NASA S69-36594

About the Authors

Roger D. Launius is senior curator in the Division of Space History at the Smithsonian Institution's National Air and Space Museum in Washington, DC, where he was division chair from 2003 to 2007. Between 1990 and 2002, he served as chief historian of the National Aeronautics and Space Administration. A graduate of Graceland College in Lamoni, IA, he received his Ph.D. from Louisiana State University, Baton Rouge, in 1982. He has written or edited more than 20 books on aerospace history, including, most recently, *Globalizing Polar Science: Reconsidering the International Polar and Geophysical Years* (Palgrave Macmillan, 2010); *Smithsonian Atlas of Space Exploration* (HarperCollins, 2009); *Robots in Space: Technology, Evolution, and Interplanetary Travel* (Johns Hopkins University Press, 2008); *Space Stations: Base Camps to the Stars* (Smithsonian Books, 2003; 2nd ed. 2009); *Imagining Space: Achievements, Predictions, Possibilities, 1950–2050* (Chronicle Books, 2001); and others. He is a fellow of the American Association for the Advancement of Science, the International Academy of Astronautics, and the American Astronautical Society, and he is an associate fellow of the American Institute of Aeronautics and Astronautics. He also served as a consultant to the Columbia Accident Investigation Board in 2003 and presented the prestigious Harmon Memorial Lecture on the history of national security space policy at the United States Air Force Academy in 2006. He is frequently consulted by the electronic and print media for his views on space issues and has been a guest commentator on National Public Radio and all the major television network news programs.

Dennis R. Jenkins has worked as a NASA contractor for the past 32 years, mostly on the Space Shuttle program in a variety of engineering and management roles. After supporting the first few Space Shuttle launches at the Kennedy Space Center (KSC), he spent 5 years activating the Vandenberg Launch Site in California before the facility was closed following the Challenger accident. Returning to KSC, he supported the recovery of the Shuttle accident and a variety of special projects. During the late 1990s, he was the ground systems lead for the X-33 program. Afterward, he managed a variety of upgrade projects at KSC. He spent 2003 on the staff of the Columbia Accident Investigation Board (CAIB), 2004 as staff to the President's commission on the Future of Human Space Flight, and 2005 as the Verville Fellow at the National Air and Space Museum (NASM), after which, he returned to KSC. In 2010, he began participating in the Orbiters on Display Working Group that planned the delivery of the Space Shuttle orbiters to their display sites at the NASM; the Intrepid Sea, Air & Space Museum; the California Science Center; and the KSC Visitor Complex.

Index

Numbers in **bold** indicate pages with illustrations.

Mercury program, 51–52, 56, 70, 130
research and writings about, 14–15
return speed and, 1
semiballistic reentries and, 47
shadowgraphs of reentry designs, **63**
skip-gliding (dynamic soaring) and, 7–8, 9,
129–130, 163
slender, streamlined designs and, 16–17,
19n40, 20, 31, **63**
Space Shuttle orbiter, 207–8, 217–19
superheating, viii, 224
temperature-sensitive paint, 175
temperatures of, 16n34, 22
X-15 aircraft, 40
X-plane research goals, 39
See also thermal protection systems (TPS)
heat pipes, 27
heat sinks
ballistic missile applications, 22–25, 26,
128, 230
concept of, viii, 26
heat absorption testing, 24, 24n53
human space flight and, 35
limitations of, viii, 35
materials for, viii, 22–25, 22nn48–49, 35
skip-gliding (dynamic soaring) and, 9
warheads (reentry vehicles) applications, 25,
26, 28, 35, 52, 59
heat transfer
flat-bottomed shapes and, 130
hypersonic flight and, 6, 8
shock layer and, 6, 279
Hello, Bastian "Buzz," 153, 153n70
Hetzel, Charles "E.P.," 109
High Speed Tunnel (Langley), 145, 149
high-temperature reusable surface insulation
(HRSI), 202–4, **203**, 211–12, 213, 216,
216n56
High Temperature Tunnel (Langley), 72
HL-10 lifting body

characteristics and design of, **133**, 148–49
development of, 133, 134–35
horizontal ladder designation, 135
hydrogen peroxide engines, 150–51
NB-52 carrier aircraft launch, 149
orbital lifting body based on, 159–160
Space Shuttle design and, 150, 151–52
supersonic flight in, 149–150
testing of, 144, 148–152, 153
thermal protection system, 150, **151**
Hoag, Peter C., 149, 151
Hoey, Bob, 42n19, 46n30, 114, 116n66,
134n23, 157n81, 159n82
Hohmann, Walter, viii, 14–15, 14n28
Hohmann Transfer Orbit, 14n28
Holmes, Brainerd, 93
hot jet research facility, **47**
hot structures
ablative thermal protection systems and,
191–92
BoMi skip-glider, 161
concept of, 9, 26
development of, 9
heat of reentry and, 165
HYWARDS program, 168
materials for, 24n53, 195–96
radiative cooling and, 9, 26
titanium, 161
X-15 aircraft, 23n52, 24n53, 26, 40–42,
168, 191–92, 195–96
human space flight
ablative thermal protection applications, ix
accidents and loss of crew, vii, 81, 209–10
beginning of program, 33
Collier's space travel articles, 36, 96
commercial piloted spacecraft, 221, 266
crew escape system, backup, 97, 98
design and shapes of vehicles for, 170
exploration beyond low-Earth orbit, 266
funding for, 266